# Marie Curie

By *Robin McKown*

Cover illustration by Ollie Cuthbertson
Cover design by Robin Fight
First published in 1959
© 2022 The Good and the Beautiful, LLC
goodandbeautiful.com

# Contents

1. Baby of the Family . . . . . . . . . . . . . . . . . . 1
2. The Governess . . . . . . . . . . . . . . . . . . . 10
3. The Sorbonne . . . . . . . . . . . . . . . . . . . 18
4. Courtship . . . . . . . . . . . . . . . . . . . . . 25
5. Young Married Couple . . . . . . . . . . . . . . 32
6. A Wooden Shack . . . . . . . . . . . . . . . . . 40
7. Fame . . . . . . . . . . . . . . . . . . . . . . . . 50
8. Catastrophe . . . . . . . . . . . . . . . . . . . . 58
9. The Long Struggle . . . . . . . . . . . . . . . . 66
10. War . . . . . . . . . . . . . . . . . . . . . . . . . 76
11. America . . . . . . . . . . . . . . . . . . . . . . 84
12. Scientists Unite . . . . . . . . . . . . . . . . . . 92
13. End of the Journey . . . . . . . . . . . . . . . . 99

# Acknowledgments

I would like to express my appreciation to those who have given me information about Marie Curie and radium (especially in its modern uses and significance), in particular Mrs. Sourya Henderson of the American Cancer Society Library; the staff of the Canadian Radium and Chemical Company; Warren M. Holm of the Radium Chemical Company; Atwood Manley of Canton, New York, who witnessed Marie Curie's visit to that town; and Otto Nathan, for permission to look at the unpublished correspondence of Albert Einstein to Marie Curie.

There is also a special debt of gratitude to Eve Curie and her publisher, Doubleday & Co., for permission to use data from her definitive biography *Marie Curie*, which in addition to being a sympathetic portrayal, is the authoritative source of innumerable facts about the author's mother.

Other sources for this story include *Pierre Curie* and *Radioactivity*, both by Marie Curie, *The Story of Chemistry* by Bernard Jaffe, *Radium in General Practice* by A. James Larkin, *Malignant Disease and Its Treatment by Radium* by Sir Stanford Cade, and *Radiation, Biology and Medicine* edited by Walter Claus and prepared for the AEC, as well as memorial and conference addresses and magazine articles in English and French.

~ Robin McKown

## CHAPTER ONE

# Baby of the Family

BACK IN 1923, THE AMERICAN publisher of Marie Curie's biography of her husband, *Pierre Curie*, requested that she write a short autobiography to add to the American edition of this work. Madame Curie protested that this could be of little interest:

> It will not be much of a book. It is such an uneventful, simple little story. I was born in Warsaw of a family of teachers. I married Pierre Curie and had two children. I have done my work in France.

Yet even during her lifetime, Marie Curie was honored and acclaimed. Twice she was awarded the Nobel Prize, and a mere listing of her scientific prizes, awards, medals, and honorary titles fills more than four pages of fine print. The biographer Emil Ludwig included her in his list of the ten greatest women in history, her portrait was placed in a temple in China as one of the "benefactors of humanity," and Albert Einstein paid her as great a compliment as any woman has ever received when he wrote at the time of her death:

> It was my good fortune to be linked with Madame

Curie through twenty years of sublime and unclouded friendship. I came to admire her human grandeur to an ever-growing degree. Her strength, her purity of will, her austerity toward herself, her objectivity, her incorruptible judgment—all these were of a kind seldom found joined in a single individual.

With her husband, Marie Curie was the discoverer of radium. Her work did not begin or end there. Einstein summed up the greatest scientific deed of her life as "proving the existence of radioactive elements and isolating them." As a result of her work, the study of physics and chemistry was revolutionized. Matter could no longer be considered inert and motionless. An atom was not a tiny solid but a galaxy with a nucleus as a sun, around which electrons, like planets, spun in orbit. Matter and energy were, in their ultimate state, one and the same thing. Inconceivable potentialities for power lurked in particles so small no microscope could make them visible. The discovery of radium and of radioactive substances, then, marked the opening of a new era—the Atomic Age.

The story of Marie Curie has the beauty of a legend. It is about a great love and a great work, and about a woman who had the qualities of a saint and yet very human emotions. It is a romantic story, sad in parts, but with the joy of achievement triumphing over sorrow.

She was the youngest of them all, a chubby little girl with short pale gold curls, big gray eyes, a stubborn mouth, and an amazingly high white forehead. Her name was Marya, but the family called her Manya, or sometimes Manyuuska, meaning "little Manya." She was born on November 7, 1867, in Warsaw, Poland.

Manya's father and mother were both teachers. The father, Vladislav Sklodovska, taught mathematics and physics at a Warsaw high school. He had been to the University of Petrograd in Russia and was very well educated. The mother taught at an exclusive private school for girls until ill health forced her to resign. She had a beautiful singing voice, and it was her hope that all her children would study music. Like many of Poland's intellectuals in this period, Manya's parents both came from families of well-to-do farmers.

There were five children in all. When Manya was four, the oldest, Zosia, was eleven; Joseph, the only boy, was nine. Bronya was three years older than Manya, and Hela was almost six. They were all blond, handsome children and bright in their studies.

Even so, Manya had startled them one day with her precocity. Bronya had been reading aloud her class lesson, stumbling through it like any youngster of seven. The baby, Manya, seized the textbook and read the passage herself, clearly and correctly. How she had learned to do this was a mystery. The parents tried to keep her away from books. It was better for a small girl to play out in the garden than to strain her eyes reading. But when visitors called, they couldn't resist asking her to recite. Manya, who was shy, would want to go and hide.

Professor Sklodovska had no laboratory at his school. In a glass case in his study at home were his instruments—glass tubes, small scales, an electroscope. Manya, wandering around his room, stopped in front of the case, fascinated.

"What are those, Papa?"

"My scientific apparatus," he explained.

Manya never grew tired staring at his "scientific apparatus."

Sometimes she heard her father speak in a low voice of Mr. Ivanov, the Russian director of the school where he taught. Even when she was very young, she understood that Mr. Ivanov belonged to the "enemy," that he was employed by the Russian government to spy on the teachers and the students alike.

Poland at this time was not a free country. Many years before, in 1807 after Napoleon was defeated, three foreign powers— Austria, Prussia, and Russia—had divided Poland between them. Warsaw, where Manya lived, was part of Russian Poland.

Twice the Polish people had organized revolts against the Russians, once in 1831 and again in 1863, four years before Manya was born. Both times the revolts had been put down, and the leaders had been hanged or sentenced to exile. But the Russians were made to realize they could not kill the people's desire for freedom by such cruel measures alone.

They ruled that only Russian would be spoken in Polish schools and churches. They sent in Russian policemen, officials, and professors. They censored newspapers and books to see that nothing was printed not in favor of Russian rule. In this way they hoped to make Russians of these stubborn Poles.

The Polish patriots decided that since they could not yet win by force of arms, they would wage a battle of wits. Secret meetings were held, and groups of intellectuals were organized, including professors and priests. They pretended to obey the Russian regulations so they could keep their positions, but secretly they taught Polish history and the Polish language to their students.

Professor Sklodovska hated Russian interference in his classes, but usually he kept his tongue in the presence of the director. Once, when Mr. Ivanov criticized a pupil for mistakes in Russian grammar, he forgot himself.

"Everyone makes mistakes sometimes, Mr. Ivanov," he protested. "Even you make mistakes in Russian fairly often."

The director was furious. Shortly afterward, Professor Sklodovska was notified that his salary was cut and that he could no longer occupy his pleasant apartment in the school building.

This was the beginning of a series of misfortunes for the Sklodovska family. The professor invested his life savings in a project of his brother-in-law's to finance a steam mill. The project failed, and the money was lost. He could never forgive himself for the harm he'd done his family by this bad investment.

To add to their now small income, they took in the professor's students as lodgers in their new apartment. These sometimes rowdy boys slept in the bedrooms, while Manya and her sisters used couches in the dining room, rising at dawn to straighten up the place before breakfast was served. The peaceful family life was a thing of the past.

Then Bronya and Zosia fell sick with typhus. Bronya recovered, but Zosia, the oldest and the merriest of them all, died of the disease in January of 1876, when Manya was eight. The blow saddened them all, and Madame Sklodovska, Manya's beautiful mother, never recovered from her grief. Her health had become increasingly poor. She had tuberculosis. Manya was nine when the family gathered around her mother's bedside for the last time. She gave to each of them words of comfort and farewell. Manya remembered her mother always for her sweet disposition, her kind heart, and her sense of duty.

Various housekeepers came to take charge of the household after that but could not substitute for a mother's care.

Manya's first school was a private one, but it too was periodically inspected by a Russian official. Many of the professor's students had difficulty mastering the Russian language, but Manya learned it with no trouble at all. When the Russian inspector, Mr. Hornberg, paid his regular visit to her school, Manya's teacher always called on her.

Manya hated it.

"Recite the Lord's Prayer," the inspector ordered her on one such occasion.

Her face white, she obeyed, speaking in a monotone the words she felt only had real meaning in the Polish tongue.

"Name the Czars who have ruled over Holy Russia since Catherine II," he demanded next.

"Catherine II, Paul I, Alexander I, Nicholas I, Alexander II . . ." she reeled off the names.

He asked more questions, and Manya answered them all to satisfy him. Finally he nodded curtly and left. He never knew that his arrival had been announced by two long rings and two short ones from the porter's bell, which had given the girls just time to clear away their Polish books and substitute their sewing.

Afterward, the teacher called Manya up to her and kissed her on the forehead. But Manya burst into tears.

In the evenings the Sklodovska children and the boarders would gather around a long table to study their lessons. None learned more easily than Manya. It was nothing for her to memorize a long poem by reading it through once or twice. She always finished her studies first, after which she might help one of the older boys to solve a difficult problem in arithmetic.

She learned fast because she knew how to concentrate. When she was studying, she would not even hear Hela, next to her, reciting her lessons aloud. One day, for a joke, the children made a scaffold of chairs around her, one resting precariously over her head. She noticed nothing until she finished her book and got up to go to bed. As the chairs crashed around her, her sisters and the boys burst into shrieks of laughter.

Manya looked at them calmly. "That's ridiculous," she said, and left the room without another word.

She was going to high school then. Originally this school had been established for German government officials, and the discipline was in the strict German tradition. It seemed that the Poles here were considered lower than anybody, yet once, an amazing thing happened. A Russian professor presented a pupil with a volume of poems by a Polish revolutionary writer. Manya could not get over it. Was it possible that even among the Russians there was some sympathy to the cause of Polish freedom? It was a wonderful thought.

The official persecution of the Polish people continued. One day Manya and her schoolmates found a young girl named Leonie weeping bitterly. They gathered around to ask what the matter was.

It was her brother, Leonie told them between sobs. He had been part of a group who were plotting against the Russian oppressors. Someone had denounced him, and he had been put in prison. The Russians were going to hang him the next morning at sunrise.

That night Manya, Hela, Bronya, and two other girls stayed with Leonie in her small room. There was not much any of them could say, but they offered her what comfort they could, bathing her face with cool towels from time to time and trying to force her to drink some hot tea. When dawn came, marking the moment of the brother's execution, all six knelt in prayer. At that moment it seemed to Manya that nothing was important except for Poland to be free again.

She graduated from her high school studies on June 12, 1883, a few months before her sixteenth birthday. Like Bronya and Joseph, she was given a gold medal, the highest award the school had to offer. After his graduation, Joseph had enrolled at the Faculty of Medicine in the University of Warsaw. But no women were allowed in the university. Bronya, who was so intelligent, was now staying at home, looking after the household and cooking and cleaning. Even at fifteen, Manya felt that to be unfair.

She had become thin and pale from her intensive studying. Her father decided she needed a vacation and shipped her out to some relatives in the country. She spent more than a year with different families, doing almost nothing but enjoying herself. It was the one carefree period of her life.

That year made her an ardent lover of nature. She went for hikes in the woods with other young people, picking wild strawberries and eating them with a large appetite. With her companions she gathered poppies and cornflowers and pinks and made them into gigantic wreaths. One uncle with whom she stayed had a stock farm with fifty thoroughbred horses. Manya, in borrowed breeches, learned to be an expert horsewoman.

Other relatives lived at the foot of the Carpathian Mountains. Much as she liked the open country, with the plains gently rolling to meet the horizon, she was enchanted with her first sight of the mountain peaks. There were excursions to valleys and to high mountain lakes with picturesque names like Eye of the Sea. Mountain climbing, following narrow trails up through fragrant fir trees and green bilberry bushes spotted with tiny alpine flowers, seemed to her the most glorious sport there could be.

For the winter months, she stayed with another uncle on the Galician frontier. He was an amiable man with three daughters about Manya's age, all rosy-cheeked and full of laughter. "How beautiful you are," they cried, clustering around their city cousin on her arrival.

Manya gazed at them in astonishment. Bronya was beautiful and so was Hela, but she had never considered herself so. Without realizing it she had passed through the awkward period of her early teens and had blossomed into an exquisite young woman with fine, delicate features crowned by a halo of golden curls that never would stay in place.

Galicia was under Austrian control, which was much less severe than the Russian rule in Warsaw. Here people could speak Polish and sing Polish songs freely, without fear of being sent to prison.

Life at her uncle's was a round of parties and entertainment. "You are going on a *kulig*," her cousins informed her one day.

"What's a *kulig*?" Manya asked.

"You will see."

The *kulig* began with a glorious sleigh ride across the snow. The girls, in peasant clothes and bundled up in blankets, huddled inside the sleigh, while their young men, also masquerading in rustic dress, rode on horseback as a mounted guard. Other sleighs of young people caught up with them, including one with four little musicians whose tunes intermingled with the rhythm of the horses' hooves on the hard snow.

They stopped in front of a darkened house and started pounding on the door. Miraculously the lights burst on, and they went inside to a feast at ladened tables, prepared much in advance. At a signal they all departed, including their hosts, and the *kulig*, growing like a snowball, continued to the next house.

All night and all the next day they flew over the snow, stopping only to eat and for a little sleep stretched out on sweet-smelling hay in someone's barn. On the second night, they stopped at the largest house in the countryside where a ball was to be held.

The musicians, who had had no more sleep than the others, launched into the liveliest of dance music. Manya, charming in her velvet jacket and puffed sleeves, found as her partner a handsome young man in a white embroidered coat. For those hours he seemed to her the prince of her dreams, though she never saw him again. They danced until eight in the morning. Manya was not tired at all.

This wonderful year of gaiety was to last little Manya for a very long time.

CHAPTER TWO

# The Governess

WHAT WAS SHE GOING TO do with her life? Her long vacation had given Manya time to reflect on that subject, but truthfully she had not done so. One fact was clear when she returned to Warsaw. She, along with her sisters and brother, was going to have to earn a living. Her father's small salary was not enough to support them all.

In Manya's absence, the family had moved to a smaller apartment and had stopped taking boarders. Once again they were alone together. The new intimacy gave Manya a chance to get better acquainted with her father.

To outsiders, Professor Sklodovska was a plump little man with a pointed beard, strict in his classes, precise and orderly in his habits. Beneath his commonplace appearance, Manya saw his kindness and wisdom. Had he lived in a free country, he would likely have risen to a high university post. But though the Russians could see to it that he never became more than a high school teacher, they could not suppress his love of learning. He kept up with all the latest scientific journals. He knew Greek and Latin thoroughly, and in addition to Polish and Russian, spoke English, French, and German. For relaxation he wrote poetry, often about the everyday events of the household, or he translated foreign classics into Polish.

Every Saturday evening the family gathered in his study to drink tea while he read to them. In their childhood he told them mythical tales or read *David Copperfield*, translating it from English into Polish with hardly a break. Now that they were older, he liked to read them the great epics of the Polish poets, from books that the Russians had banned.

He could not give a dowry to his daughters or send them to a foreign university, but he could and did share with them the riches of his mind.

That winter Manya and Bronya did some private tutoring to small children of prosperous families. It was a thankless task. They had to travel all over Warsaw in the snow and sleet. The pay was low, and often the lady of the house would "forget" to pay them at all.

Manya found another release for her energies through a group of young people who called themselves the "floating university." Like herself, they were ardent patriots. Their creed was that the hope of Poland lay in strengthening the intellectual and moral force of the nation—that one cannot hope to have a better world without improving the individuals of that world. Always in the greatest secrecy they met at each other's houses to discuss the future of Poland, and they held classes in the evenings to teach subjects not permitted in the Russian-controlled schools. Thus after her paid tutoring, Manya gave lessons in Polish history to a group of underpaid women employees of a dressmaking house.

This activity only increased her desire to learn more herself. Through a cousin, she gained access to a small laboratory. Evenings and Sundays she went there, usually by herself, to try out experiments described in books of chemistry and physics. Sometimes her experiments turned out successfully. Just as often they failed.

There were so many exciting things in the world of knowledge. In her room late at night, she read Dostoevski and other great writers, as well as books on philosophy, poetry, biology, and science. Everything interested her. But more and more she realized that her deepest enthusiasm was for physics and chemistry.

By this time her brother Joseph was nearly finished with his university studies and would soon be launched as a doctor of medicine. Pretty and vivacious, Hela was doing all right too. She was the only one of the children who had continued with her music after their mother's death. She was torn between a career as a singer and the teaching profession and was talented in both. In the meantime she had to fight off dozens of proposals of marriage.

It was different with Bronya. Like Joseph, Bronya wanted to be a doctor, but as long as she stayed in Poland, it was out of the question. It seemed a terrible waste to Manya that her beloved older sister must fritter away her time giving lessons to backward children who didn't appreciate her in the least. That she had to do the same was a minor matter.

She brooded over Bronya's fate, searching for the solution. Eventually she found one. Bronya had saved enough money for one year. The medical course took five years. Manya herself would supply the funds for the remaining four years by working as a governess. That way she would not have to pay for her board. Nearly all her salary could go to Bronya.

It was not easy to convince Bronya of the beauty of such a practical plan. Why should her younger sister condemn herself to such drudgery? Bronya asked. Manya was just as talented as she was, if not more. Why should she make the sacrifice?

"Because you are twenty and I am seventeen," Manya replied firmly. "I have plenty of time." She added, "Besides, when you are practicing medicine, you can help me."

It was this afterthought that made Bronya decide to accept little Manya's generous offer.

## THE GOVERNESS

Her first post as governess was in Warsaw. The family she worked for had all the worst faults of the newly rich. They spent money lavishly for show, but scolded their servants for using too much oil in the lamps and kept creditors waiting six months or so for payment. They liked to employ French phrases in their conversation to make an impression, but always mispronounced them. The same guests they greeted with flattery they slandered behind their backs. Manya's first view into the lives of the upper classes filled her with horror.

She had another reason for looking for another post. As long as she still lived in Warsaw, she was tempted to spend more money than she could afford, if Bronya was to be provided for. Still, the day she took the train that would carry her from her family and Warsaw to live among strangers was one of the saddest she had known.

The place where the employment agency sent her was called Szczuki, and her only hope was that the countryside would be pretty. When she woke up the first morning after her long journey, she looked out her window to see dreary factory buildings with grimy chimneys sending forth clouds of smoke. There were no trees or green meadows in sight—only endless fields of sugar beets. Mr. Z, her new employer, owned some two hundred acres of these beets and was part owner of the adjoining sugar beet factory.

He was very nice to her and so was his wife, who treated her like her own daughter. What a blessing it was for them, they said, to find a governess who was so pretty and talented and well mannered. Often after her classes were finished for the day, Mr. Z would insist Manya go for walks with him, when he would explain in great detail to her the distribution of the crops in the fields and the modern techniques of agriculture. It delighted him to find a young woman who was not at all frivolous and who grasped these subjects as quickly as any man. Nor did Manya have to pretend to be interested. Agriculture was a new subject to her, and any new subject held her attention.

There were three small children in the household and an older daughter named Bronka, of Manya's age, who soon became her friend as well as her pupil. She and her parents tried to include the governess in neighborhood parties and dinners. After the first, Manya always tried to find excuses. She found the affairs boring and the conversation trivial and shallow. But she could not resist invitations to go horseback riding or skating and on sleigh rides in the winter. Sometimes on such rides, they could hardly see the road for the snow, and the passengers called out to the driver, "Look out for the ditch. You are going straight into it."

"Never fear," the driver called back. And a moment later, over they went, tumbling and laughing, into the snow.

One winter day, when the snow was high in the fields, Bronka and the children of the household built a marvelous snow house where they could sit and look from their retreat across the rose-tinted plains.

Sometimes when Manya was out walking, she would see the peasant children of the neighborhood, poorly dressed and with unkempt hair, and occasionally she stopped and talked to them. It was a shock to find out they could neither read nor write. At most they had gone to school just long enough to learn the Russian alphabet. How wonderful it would be to teach them the Polish language and history!

She talked her idea over with Bronka, who was enthusiastic.

"You know if we are denounced, we will be sent to Siberia," Manya warned her.

The thought of danger made the project more exciting in Bronka's eyes. She got permission from her parents for "Mademoiselle Manya's school," and the two of them went calling on the peasants and farmers and factory workers to explain what they had in mind. After that, Manya's room, on the first floor and with a separate entrance, was open to the young sons and daughters of the poor people of the neighborhood on Wednesday and Saturday afternoons. They came singly and in twos or threes, and finally there were eighteen in all. They stared awestruck at the blond young lady in the plain black dress who was so kind and patient with them. When the big black letters she drew for them on white paper finally made words they could read, they fairly burst with pride.

Eighteen Polish children rescued from ignorance! It was a very little thing, Manya told herself. But she was as proud as they were.

The oldest son of the family, a young man named Casimir, returned from the University of Warsaw for the holidays. Almost overnight he fell in love with this unusual governess, who was lovely to look at, who could dance, skate, ride horseback, and even compose poetry. He was handsome and charming. Manya returned his affection. They decided to get married.

Casimir went to his parents to get their consent. Neither he nor Manya expected any objections. Had they not always treated her as a member of the family? To their amazement Mr. and Mrs. Z said no, very definitely, to the idea of their son marrying Mademoiselle Marya. She was, after all, they pointed out, a governess. It was not fitting for their son to marry a governess. They would not even consider it.

Manya was bitterly disillusioned. The friendship the family had offered her had been false. She was even more disappointed in Casimir's reaction. Instead of standing up to them, he pleaded with Manya to wait. Everything would turn out all right in the end. She didn't believe him.

After Casimir returned to Warsaw, life at the Zs' became strained and difficult. But she couldn't quit, not just yet. Bronya, who was making good progress at the Sorbonne, had exhausted her own savings and was dependent on Manya and whatever their father could spare.

Manya stayed on three years in all, the last part of which was a painful ordeal. When she thought of her own early ambitions, it was like opening a deep wound. Sometimes it seemed to her she would always be living away from home, teaching other people's children from her own woefully inadequate fund of knowledge.

The time came when Professor Sklodovska was retired on a pension. He started at once to look for a new position and found one as director of a reform school outside of Warsaw. It wasn't his own kind of work, but the pay was better than he had earned as a professor. He wrote Manya that she could save her money now for herself; between his pension and his new salary, he could pay Bronya's expenses in Paris.

Shortly after that Manya left the beet fields and Szczuki forever, to accept a new post in Warsaw, where she could once more be near her father. Her new mistress was beautiful and elegant. She wore Paris-designed gowns, jewels, and furs in abundance, and the most expensive perfumes. Manya wouldn't have been normal if she had not found some pleasure in being near all this luxury. But she didn't envy her new employers. She had discovered in her years as a governess that all the money in the world cannot make up for empty minds.

An exciting letter came from Bronya. She had passed all her examinations except the final one, and when she passed that, she would be a full-fledged doctor, like Joseph. More than that, she was going to get married. Her fiancé was a doctor too, named Casimir Dluski. As soon as they set up housekeeping, they wanted Manya to come live with them. It would cost her nothing, and she could study at the Sorbonne.

Though it was what she had dreamed of for so long, Manya at first refused. As soon as she finished her present post, she intended to keep house for her father, she wrote Bronya. He was old and needed her now. There was another reason for her decision that she did not admit to anyone, even to herself. She was still attached to her own Casimir and could not bear the thought of being so far away from him.

In the fall she took a vacation in her beloved Carpathian

Mountains. Casimir met her there, and as they hiked up into the hills, he once more went over all the obstacles to their marriage. What would Manya suggest he do?

Suddenly Manya had had enough. "If you can't see how to solve our situation, it is not for me to teach you," she snapped at him.

That night she made up her mind—she was through forever with such sentimental nonsense. Other women might marry and have children, but that was not for her. Her true love was science. It was to science that she would give her life. Before she could change her mind, she wrote to Bronya saying that she would accept her offer, if Bronya still wanted her. She knew that her father would miss her, but she also knew that he would approve her decision.

Bronya replied promptly to say that Manya's room was ready for her.

The next days were filled with furious preparations. She had to get a passport. She had to send ahead certain household items that Bronya had told her she would need—a mattress, blanket, sheets, and towels. For her few personal belongings, she bought a sturdy wooden trunk.

When all this was taken care of, she had barely enough money for her fare. Third-class passage was not very good, but across Germany she found out there was a still cheaper way to go—in fourth-class carriages. These were as bare as freight cars. She even had to provide herself with a small folding chair to have a place to sit down. But the discomfort of the three-day voyage meant nothing to her. In addition to her chair, she carried along food enough to last her for the journey and a few books. With these meager supplies, Manya, at twenty-three, set forth for her new life as a university student.

CHAPTER THREE

# The Sorbonne

BRONYA WAS A BORN HOMEMAKER. The Dluski flat, on the outskirts of Paris near the park of Buttes-Chaumont in La Villette, was charmingly furnished with draped curtains, curved Venetian furniture, and an upright piano. Meals were delicious. From the beginning Manya was good friends with Bronya's young husband, who was good-looking, charming, and brilliant.

Casimir Dluski was an exile from Poland. While he was studying in Petrograd, he had been accused of taking part in a plot to assassinate the Russian Czar and had had to flee from Russia. The Czar's police had lodged reports against him with the Paris police, which prevented him from getting his French citizenship papers but did not interfere with his zest for work or for pleasure.

They had fitted up one room of their apartment as an office, and part of the day Casimir used it for his patients—mostly butchers and workers from the nearby slaughterhouses. At other hours, Bronya treated the butchers' wives there. In the evenings they loved to entertain other Polish exiles or go to the theater.

Casimir always insisted that their little sister must share their amusements.

One evening he informed Manya she was going to a concert with them. It did her no good to protest that she couldn't leave her studies. He had tickets for all of them, he said. A Polish pianist was playing. He was marvelous, but he was unknown, and the hall would probably be half empty. They must go and applaud until their hands hurt.

With some reluctance, for she really wanted to stay home and study, Manya let herself be persuaded. It was one time she wasn't sorry. The pianist was a tall, thin young man with flaming red hair, dressed in a threadbare coat. But his playing was beautiful. Their fellow countryman Frederic Chopin took on a new beauty as the long fingers captured the composer's melodies from the piano keys. Manya, who adored music as her mother had done, felt herself carried to the stars.

Afterward, the shabby, hungry-looking young man came to their house for tea with his ravishing fiancée, Madame Gorski. Manya remembered his name—Ignace Paderewski; one day he was to be recognized as one of the world's greatest pianists and to become prime minister of a liberated Poland.

Everything about Paris enchanted Manya from the first moment when she got off her train at the grimy Gare du Nord after her terribly tiring journey. She was impressed not so much with the beauty of the city—Warsaw too had its own charm—but by the feeling of freedom that seemed to be part of the air one breathed. It was wonderful to hear people talking their own language on the streets without fear.

She could have spent weeks sightseeing, but she didn't. Right after her arrival, she set forth for the University of the Sorbonne, the place that had become to her an enchanted palace where all her dreams would come true.

The Sorbonne had been founded originally in the thirteenth century by Robert de Sorbon, chaplain of Louis IX, for poor theological students. Since then, through its doors had passed the most brilliant minds of all Europe. When Manya first saw it, the gray buildings built in the time of Richelieu were being renovated, and the modern workmen's shanties and scaffolding contrasted strangely with the ancient stone. She hardly noticed its external appearance.

What first caught her attention was a white poster on the walls. FRENCH REPUBLIC—FACULTY OF SCIENCES—FIRST QUARTER, it read. *Courses will begin at the Sorbonne on November 3, 1891.* Beneath that was a list of the various courses that her small savings would give her the right to attend. It still seemed too good to be true.

She did not sign her registration card with the childish name of Manya, nor even with the more dignified "Marya." She was in France now, so she wrote her name in the French way: *Marie Sklodovska.*

Even in Paris where there were no laws against it, few women wished to specialize in physics and mathematics. It was no wonder that Marie's fellow students were curious about her. What woman would want to devote herself to science unless she were so unattractive she could not have another life? Or unless of course she hoped to find herself a husband in this man's world. Yet this attractive young woman with the soft pale blond hair and the plain serviceable dress was obviously not aware that the young male students existed. She seemed to move as in a dream.

"Who is she?" one student asked another, as he watched her walk down the corridors.

"A foreigner with an impossible name," he was told. "She is always in the first row at the physics course."

No, Marie was not in the slightest interested in flirting with the young students. She did wish with all her heart for the wisdom lodged in the brains of the older professors who taught in the Faculty of Science.

That first session was very hard for her. In spite of her diligent study of the French language, she was lost in some of the lectures where the professors talked very fast in technical terms. Nor had her home study given her the background that the other students had, particularly in mathematics. She found great gaps in her learning and realized she would have to work harder than all the others to catch up.

One of her favorite professors was Dr. Paul Appell, an Alsatian by birth, a grave man with a squarely cut beard. Mathematics to him was not merely a matter of adding or subtracting, or of theorems and equations. It was the key to the laws that govern the universe.

"I take the sun, and I throw it," he said once, to illustrate some point.

Marie heard him with ecstasy. It was as though he were saying that through science man could master all the secrets of nature. Now she knew why she must be a scientist herself. To scientists was given the mighty glory of finding the reason and logic behind the creation of the world. What did the drudgery that science demanded of its followers matter, compared with the rewards?

The Dluski flat was on the other side of Paris from the Sorbonne. To get to the university, Marie had to take two different buses—riding on top because it cost less. She didn't mind the ride, but she resented the waste of time. No one could do much studying on top of a Paris bus.

The other handicap to living with her sister and her delightful brother-in-law was that she simply couldn't work evenings. There were always people around, always interruptions. She came to the conclusion that if she were to accomplish what she had come for, she must live alone.

Bronya and Casimir protested vigorously, but when they realized nothing would change her mind, they helped her move her things to a small room at 3 Rue Flatters, in the Latin Quarter, which she had rented because from there she could walk to the university in twenty minutes and to the chemistry laboratories in fifteen.

She stayed at this address only a short while. The rooming house was filled with the chatter and clatter of other students and young officers. She still didn't have the privacy and quiet she felt were essential.

Her next move was to a still cheaper place, a tiny attic room. It had a skylight but no windows, and it was unheated. She furnished it with the bare essentials—an iron bed, a table, a kitchen chair, an oil lamp, and a small alcohol heater on which to prepare meals. When Bronya and Casimir came to see her, they had to sit on the wooden trunk she had brought from Poland.

She had a washbasin and a pitcher that she filled with tap water from the landing. Two sacks of coal were sufficient for the winter; she carried them up the six flights of stairs herself, bucket by bucket. For light, the skylight was enough until night fell. Then she went to the Sainte-Genevieve Library, where it was warm and where students like herself were allowed to study until ten o'clock closing hour. Thus she needed to light her own lamp only for the studying she did after ten at night—which usually lasted until two or three in the morning.

The budget she made up for herself allowed her just a hundred francs a month for everything—rent, clothes, paper and books, and food. It was very little, though other students, Bronya, for instance, had managed on no more. But Bronya had lived with two or three other girls, and they had cooked their meals together.

Marie didn't want to share her room, for that would have meant sharing her precious time as well. Unlike Bronya, who had been housekeeper in their father's home, she knew nothing about cooking. In fact her Polish friends said of her, "Mademoiselle Sklodovska doesn't even know what soup is made of."

Nothing seemed of less importance at this period of her life than food. She found she could live quite well on buttered bread and tea, supplementing this foolish diet once in a while with an egg or a morsel of chocolate or some fruit. It never occurred to her to buy a chop from the butcher and cook it.

When the Dluskis commented she looked thin and pale, she told them she had been working hard, which was true. After several months without a real meal, she sometimes lost consciousness just before she went to bed. The sign of weakness made her ashamed.

One day on her way home from the university, she fainted on the street near some of her fellow students. Someone summoned the Dluskis. Two hours later, Casimir bounded up the steps to her room where she was, against her will and desire, resting on her cot.

He took one good look at her. Then, over her objections, he began searching the room. The nearest to food he found was a package of tea.

"What did you eat today?" he demanded.

She stalled. "I don't remember."

He finally got her to confess that since the day before she had had nothing but a few cherries and a bunch of radishes. She had worked until three in the morning and had slept four hours.

"There's only one thing the matter with you," he informed her angrily. "We're going to do something about that."

He made her gather together the books and clothes she would need for the next week and come home with him. Once there, he spoke a few words to Bronya, who dashed for the kitchen. In almost no time, Manya, under Casimir's stern gaze, was eating a big steak, cooked rare, with a heaping plate of crisply fried potatoes.

One could almost see the roses coming back in her cheeks the next week, what with Bronya's cooking and Casimir making her go to bed by eleven or earlier. But when the week was over, she returned to her attic room and to her old regime of eating little and sleeping less. Once again she regulated her life by her small budget, on which the necessity of buying a new pair of shoes assumed the proportions of a calamity. As for other clothes, she simply didn't buy any.

Fortunately she had one household skill—sewing—which she had learned in grade school. When her brain was too fatigued for study, she washed and mended and patched the garments she had brought from Poland. She wasn't the best-dressed young lady in Paris, but she managed, in spite of everything, to look well groomed and neat.

As the months passed, she not only caught up with her fellow students, but she surpassed them. She didn't neglect her French, either, and soon she had no further difficulty understanding the lectures, no matter how fast or how technically her professors talked. She almost completely lost her Polish accent and learned to write in her adopted language with elegance and precision.

More than anything else she did, she loved working in the school laboratory. The atmosphere of silence and concentration was perfectly suited to her character. In her linen smock, she could stand for hours before her table, watching some strange bubbling concoction with the intensity of an ancient alchemist awaiting a miracle. The young man who shared the laboratory facilities with her soon learned better than to try to talk to her at such times. For Marie, then as always, a laboratory was an almost sacred place.

In July of 1893, she took her examinations and not only passed them but was first in her class. She had now a Master in Physics. She returned to Poland to visit her father for the summer, where her relatives took turns in trying to fatten her up. Then she was back in Paris again, in another garret room, to try for her second degree—in mathematics.

At a critical moment, when her funds were almost exhausted, she was granted the Warsaw "Alexandrovitch Scholarship"—a sum of enough rubles to continue her studies for another fifteen months with her low standard of living. It seemed too good to be true.

CHAPTER FOUR

# *Courtship*

EARLY IN 1894 MARIE WAS assigned her first original piece of research. It was a study on the magnetic properties of various steels, requested by the Society for the Encouragement of National Industry. She undertook the assignment with outward calm and inward joy. The trouble was that the only place she was allowed to work was in the overcrowded laboratory of her professor, Gabriel Lippmann. It had simply no space for her and her metals and equipment.

She confided her problem to some Polish friends, Professor Kovalski and his wife, who were in Paris for a visit.

"I know someone who might be able to help you," Kovalski said. "He is a teacher at the School of Physics and Chemistry. Perhaps you have heard of him. His name is Pierre Curie."

Marie listened attentively as he went on to tell her about this young scientist, Marie's senior by only nine years, who had already made a reputation for himself. In the course of a study he was making of crystals with his older brother, Jacques, he had discovered piezoelectricity, the name given to the phenomenon wherein mechanical pressure exerted upon certain crystals resulted in the production of an electrical charge along the surface of the crystal, and where, conversely, the crystal could be distorted by the application of an electric charge. Later the two brothers had invented a "piezoelectric quartz," an apparatus that could measure with precision small quantities of electricity and electrical currents of low intensity.

Sometime before, Jacques had been appointed head lecturer in Mineralogy at Montpellier University in the South of France, and Pierre had continued his experiments alone. In the course of his work on crystals, he had invented an ultra-sensitive scientific scale, since named the Curie Scale. He had also done research on the effects of temperature on magnetism, which resulted in his discovering an important fundamental rule, called after him, "Curie's Law."

It struck Marie as strange that her Polish friends should know all about Pierre Curie's achievements, while she, who lived in Paris, had never heard of him.

That was the way things happened, Kovalski said, shaking his head. Men of talent were often recognized in foreign countries before their own countrymen appreciated them. For instance, Lord Kelvin, the illustrious British scientist, had been so impressed by Pierre Curie's invention that he had made a trip to Paris to meet him. But what recognition did he have in France? None. At the school, where he was now director of the laboratory, he earned no more than an average factory hand.

Marie resolved then and there not to bother a scientist of such distinction with her small problem. Nonetheless, she was curious to meet him. The Kovalskis arranged for them both to come to their room one evening.

When Marie arrived, Pierre was already there, standing in the window recess near a door leading to the balcony. Though she knew he was thirty-five, he looked much younger. He was tall and thin with chestnut-colored hair, and his features were regular beneath his rough pointed beard. His clothes hung loosely, but there was an unconscious elegance about his carriage. But most of all, she was struck by his clear, penetrating, kindly eyes.

At first he seemed embarrassed and spoke very little, like a man unaccustomed to the presence of women. Marie asked him a question about his work with crystals, and he gave her a startled look as though she had intruded into a domain where women did not belong. Gradually he seemed to accept the fact that Marie was no more concerned with social chitchat than he was, and he began to relax. Stimulated by her interest, he discoursed on his latest research, which had developed from his study of crystals—the effects of symmetry on the phenomena of physics.

For Marie it was a truly exciting evening. Rarely had she met a man with such an excellent mind and such a clear way of expressing himself, once he was encouraged to talk.

They met casually a few times after that at sessions of the Physics Society and other scientific meetings. One day he unexpectedly gave her a present. Not flowers or chocolates, but something she valued far more: a scientific treatise he had written entitled "On Symmetry in Physical Phenomena: Symmetry of an Electric Field and of a Magnetic Field." On the flyleaf he had inscribed it in his awkward handwriting: *To Mlle. Sklodovska, with the respect and friendship of the author, P. Curie.*

She didn't realize what a compliment he was paying her. Pierre Curie was known as a woman hater. In his earlier years he, like Marie, had had an unhappy love affair. He too had vowed to think henceforth only of the pure joys of science. In his diary he had written, *Women of genius are rare.*

Sometime later he asked her if he could visit her. She received him with dignity in her poor room. He was not shocked by the obvious signs of poverty. Indeed, he seemed to understand that it was right for a dedicated scientist to live in such a barren setting.

They had been on the basis of good friends for some months

when he invited her to visit his parents, who lived in Sceaux, a small town in the woods not far from Paris. Even then she did not guess that his invitation meant he had become very serious about this young Polish student.

His parents lived on the Rue des Sablons (many years later renamed Rue Pierre Curie) in an old house half concealed behind the foliage of an attractive garden. Dr. Eugène Curie, Pierre's father, was a tall man with exceedingly brilliant blue eyes and a childlike expression. He was a practicing doctor who had never given up his research. He had investigated inoculation for tuberculosis at a period when the bacterial nature of this malady was not yet established.

He was a fighter too. During the Revolution of 1848, he had taken care of the wounded and had himself been wounded, when his jaw was shattered by a bullet. During the cholera epidemic, he had stayed on to take care of the sick in a section of Paris from which his colleagues had fled.

Madame Curie, the mother of the family, was the daughter of a dress manufacturer who had gone bankrupt during the Revolution of 1848. She was a warm, comfortable person, plump and smiling in spite of poor health, and a marvelous cook.

Pierre had not attended elementary school but had been instructed first by his mother and later by his father and older brother, Jacques. His favorite haunt was the nearby woods, where he gathered plants and small animals for his father's research. The creatures of the ponds—frogs, salamanders, dragonflies—were the companions of his childhood.

As in the Sklodovska family, the Curies were held together by a love for things of the mind and spirit. Marie quickly felt at home with them.

Two notable things happened to her in that year of 1894. She got her Master of Mathematics degree after passing examinations second in her class—and Pierre Curie asked her to marry him.

She refused him. Much as she liked him, marriage was out of the question, she said. She must return to Poland when she had finished her education. That was where she belonged.

Pierre respected her loyalty to her country, but more than he had ever wanted anything, he wanted her in Paris. Science had no boundaries, he argued. It was her duty to stay in France where there was still so much for her to learn. With the greatest reluctance, he saw her leave to spend at least the summer in Poland with her family.

All that summer his letters followed her. They were long letters written from the heart. Though they were not technically love letters—he signed himself always "your devoted friend"— the feeling he had for her was obvious in every line.

She told herself it was not because of his entreaties that she returned in the fall for another year at the Sorbonne, and in part this was true. He was right in that the university still had a lot to teach her. Nevertheless, she could not hide the pleasure she felt in seeing him again.

He resumed his entreaties, for he was not a man to give up easily now that he had decided there was no other woman in the world like Marie. Craftily he pointed out how much they could accomplish if they worked together for the cause of science. When even that approach didn't win her, he offered to go to Poland and live there with her.

This was a sacrifice she couldn't accept. To drag him away from his work into a country where he couldn't speak the language, a country where intellectuals were regarded with suspicion? Ah, no! That she would never do to a man she cared for as much as she did for him. She had to admit that she cared. Her heart had finally been won over completely by his tender and respectful courtship.

They were married on July 26, 1895, in Sceaux. It was a simple wedding, at which Marie wore a serviceable navy-blue suit. Her future mother-in-law had given it to her, but Marie had selected the material herself—she wanted something she could wear later to go to the laboratory.

Following the ceremony there was a reception in the garden. Professor Sklodovska and Hela had come from Warsaw, and Bronya and Casimir, of course, were there. Otherwise, the only guests were a few close friends from the university. Dr. Curie and Professor Sklodovska liked each other at once.

"You will have a daughter worthy of affection in Marya," the professor told the father of Marie's husband. "Since she came into the world, she has never caused me any sorrow."

Marie had received a sum of money from a relative as a wedding present, and she spent it on what seemed the wildest extravagance. She bought two bicycles. It was on these bicycles that they took their honeymoon.

They had no fixed itinerary, but wandered at random through the lovely provinces surrounding Paris. Their baggage was light—a few practical clothes and two long rubber capes to protect them from showers. When they were hungry, they stopped in some shady glade and ate bread and cheese and fruit. At night they always managed to find some country inn, where their hosts would bring out plates of hot and filling soup.

Often they parked their bicycles and went off for a hike through the woods, Pierre striding ahead, Marie following him, bareheaded, dressed in a white blouse and a skirt that, contrary to fashion, was shortened to ankle length. They must have looked much more like a young peasant couple than like scientists.

Pierre loved the outdoors as much as Marie did, but on this trip he became her teacher, for he knew more than she did about nature lore. He was familiar with every plant and tree, and at a pool surrounded with reeds, he pointed out to her all the small creatures who made their home there. One day he startled her by placing a frog in her hand. When she protested, he said to her in all seriousness, "But a frog is a most amusing animal. Let us watch him." Later he risked a ducking when he climbed out on a fallen tree trunk to gather her masses of yellow irises and pale waterlilies.

Later in the summer, they met Bronya and Casimir at a lovely farm in Chantilly, in the midst of a forest. Professor Sklodovska and Hela, who had not yet returned to Poland, joined them, and often Dr. Curie and his wife came up to visit. Marie was delighted to see how well Pierre got along with her father. To please her, he embarked on the most difficult scholastic feat of his career to date—the mastery of the Polish language.

They stretched their wonderful holidays to October. Then they were ready to return to Paris and their chosen work.

CHAPTER FIVE

# Young Married Couple

THE NICEST THING ABOUT THEIR tiny apartment on 24 Rue de la Glacière was that it overlooked a garden. Marie did not try to make the three small rooms cozy and homelike as Bronya might have done. Their few pieces of furniture were castoffs from their relatives. The biggest room was their study. It was furnished only with a long table and a straight-backed chair at each end, one for Marie, one for Pierre. If visitors intruded, there was no place for them to sit down. The walls and floors were bare and the windows uncurtained. The one bright spot was the flowers Marie always kept on the table among their books and papers. Flowers were cheap and were sold at open stalls everywhere in the streets of Paris.

They worked at their table every evening after supper. Pierre prepared his class lectures on crystallography (the science of crystals) and on electricity. He was planning a new course in this latter subject, for engineers, which would later be the most complete and modern of its type in France. He was also, on his own, doing research on the growth of crystals.

Marie was studying for the certificate that would give her the right to teach physics in a school for girls. She worked as intently as she had in her little attic room, but there was a big difference. She was not alone. She and Pierre might be too preoccupied to talk, but they could at least glance across the table from time to time and exchange a smile or a loving look.

The better she knew him, the more attached to him she grew. What a wonderful person he was—tender and wise and unselfish. He never lost his temper. "Getting angry is not one of my strong points," he said once, almost apologetically. That didn't mean he wouldn't fight for what he believed. Once he had decided something was right, nothing could change him.

Since he was so much older than she, Marie looked up to him as her guide. She considered his work far more valuable than hers and consulted him on everything she did. Pierre, on the other hand, found her quick mind a constant source of inspiration. He was always saying he still couldn't believe his good fortune in finding a woman who could grasp the principles of physics with the ease she did.

It was as though they were two parts of a chemical compound, meant to be united. Their thoughts, their dreams, their ambitions seemed to flow from a single source. They were supremely happy.

They didn't have much money. Pierre's salary at the School of Physics was 500 francs a month, and there were few extras. Marie couldn't expect her husband to live on cherries and tea as she had done. Nor did she want Pierre's mother, who was such an excellent cook, to think that the Polish girl her son had married could do nothing. She set herself to learn the secrets of the kitchen with the same application she used in her studies.

Of course there were a few burnt roasts and other such mishaps at first. Pierre, who never knew what he was eating anyway, didn't notice. As Marie's skill increased, she invented dishes that would cook themselves while she was away at her classes. She did her marketing early in the morning, and before the end of their first year of marriage, no one would have recognized the "Mademoiselle Sklodovska who doesn't know what soup is made of."

She passed her examinations, first in her class again, in July. The Curies packed their bags, shipped their bicycles, and were off for Auvergne, in the mountainous region of South Central France. A sunny morning found them crossing the fresh green fields of Aubrac. By moonlight they climbed to a high plateau where a herd of cows stared at them with great sad eyes. At dusk they stood by the gorges of the Truyère River, while a haunting folk song echoed from a boat in the distance. There were other lovely memories. As they had done on their honeymoon, they traveled from place to place, never planning in advance where they would spend the night.

Whenever in the next years anyone told Marie that she and Pierre worked too hard, she would say, "That isn't so. We take vacations every summer." It was true. Sometimes, too, they found time in the midst of their busy work year to take a Sunday off to go out in the woods and meadows around Paris. Pierre loved it and was able to share with Marie all the delightful places he had explored on his own before he met her.

The Forest of Compiègne charmed them in the spring, with its moss and green foliage and periwinkles and anemones. On the banks of the Loire, at the border of the Forest of Fontainebleau, they gathered water buttercups to take back to Paris with them. Often, too, they went for weekends to Pierre's parents in Sceaux, where Pierre's old room was always waiting for them.

It was perhaps these brief interludes that saved them from physical collapse during the fearfully hard work program they set for themselves.

In the second year of their married life, Marie finished and published her investigation of magnetic properties of steel, the study that had first brought her and Pierre together, and received a small payment for it. That year, too, on September 12, her first child was born, an exceptionally beautiful baby girl whom they named Irène.

Shortly after that Pierre's mother died, and old Dr. Curie came to live with them. The apartment was too small for four. They moved to a house at 108 Boulevard Kellermann, near the park of Montsouris, by the fortifications of old Paris. Many women are torn as to whether they should have a family or a career. Marie never had any doubts about this. She wanted to do both jobs, and she wanted to do them well. Dr. Curie's presence made this decision easier for her. For he became at once the self-appointed nurse and caretaker for little Irène.

Though Pierre detested pointless social gatherings as much as Marie did, he had close friends among his colleagues, and his students adored him. Sometimes they would all sit out in the garden and talk about the new theories in physics, while Marie would sit and listen as she sewed baby clothes for her small daughter.

One of the scientific events they discussed with excitement was the new discovery of a Bavarian professor of physics named William Roentgen. Roentgen had been investigating fluorescence in zinc sulfide. Fluorescence is the quality certain minerals have of glowing when they are subjected to light. For his experiments he used a Geissler tube, a tube filled with rarefied gas, invented by a German glassblower named Heinrich Geissler, which lighted up when an electric discharge passed through it.

One day he had wrapped a lighted Geissler tube in black paper and was astounded to see that the zinc-sulfide screen near it began to glow, even without being directly subjected to light rays. He named his find "X-rays" because they were like rays of light but were not light. That is, the "X" stood for the unknown qualities of these rays.

It had been found that these X-rays could penetrate wood and hard rubber, though they were stopped by heavy substances such as lead. They could also penetrate flesh and thus be used to photograph the skeletal structure of the body, though few realized yet what they would mean to surgery and medicine. Their discovery, however, had inspired other scientists to see if similar rays might be produced by other fluorescent substances.

The scientist Henri Becquerel, for instance, had been testing the fluorescent qualities of uranium salts, and in the course of his experiments had come across another unexpected phenomenon. He had happened to place a piece of uranium ore on a photographic plate covered with black paper. The next day there had been an impression on the photographic plate, as though it had been exposed to light. It seemed unbelievable. He repeated the experiment at other times, using uranium that had been kept in darkness for months. Invariably the photographic plate showed an impression. The only conclusion was that there must be unseen rays emanating from the uranium, which were not dependent on outside light!

Thus Henri Becquerel was the discoverer of the quality in certain elements to which Marie Curie later gave the name of *radioactivity*.

She was looking at that time for a subject for her doctor's thesis. As she listened to Pierre and his colleagues, it struck her she would like to learn more about the radiation of uranium, starting her experiments where Becquerel had left off. It would be something new, something no one else had done before her.

She talked to Pierre about it. "I'm going to find out the secret of these strange rays. That will be the subject of my thesis."

"You might just be able to do it," he said thoughtfully.

Neither of them guessed what a superhuman task she had set for herself.

The first problem was to find a laboratory where she could work. This wasn't as easy as it sounds, and all the scouting around that she and Pierre did met with failure. Finally Mr. Schulzenberger, the director of the School of Physics and Chemistry where Pierre taught, gave her permission to use a damp and unheated office on the school's ground floor on the Rue Lhomond.

Here Marie assembled the equipment she would need—an electrometer; a piezoelectric quartz invented by Pierre and his brother, Jacques; and a chamber of ionization. The small room was very crowded indeed when she had everything in place.

Uranium rays had the power to make of air a conductor of electricity, and this was the key as to how they could be detected and measured. Marie started out by pulverizing uranium ores and spreading them in uniform layers on plates of the same diameter that could be inserted into the ionization chamber. Then, using the piezoelectric quartz method, she measured the saturation current produced in the chamber between the plates.

Within a few weeks, she learned that the intensity of radiation increased with the uranium content of her ores and did not change with light or heat or cold or chemical treatment. This could only mean that radiation was part of the atomic structure of the element of uranium, which was a very interesting conception indeed.

Were there other elements that had rays like those in uranium? Blithely she set to work to test all the elements that were known at that time. This was the only way, for science does not permit guesswork. When the long series of tests were over, she had found one other element, thorium, which like uranium was radioactive.

She couldn't always work with elements in pure form. Sometimes they were mixed in minerals. When she tested the uranium ore pitchblende, she discovered a curious thing. Pitchblende was about four times more radioactive than pure uranium.

It didn't make sense. It was as though one added two and two and got eight. She repeated her experiment over and over again. Always the results showed that pitchblende was far more radioactive than uranium in its pure form. The same proved true with ores containing thorium.

Since she had made certain that no other known elements were radioactive, there could be only one explanation. In addition to uranium, pitchblende must contain an *unknown* element, one that was powerfully radioactive. She told Professor Gabriel Lippmann from the Sorbonne about her findings. He was so impressed, he insisted on making a report to the Academy of Science, which was duly published in their official *Proceedings* on April 12, 1898. The small notice was sufficient to make a stir in scientific circles all over Europe. In Germany, Professor Runge of Göttingen commented with light irony: "Nature is getting more disorderly every day."

Pierre realized that his wife was on the verge of something very important to science, and he could not let her go on alone. Temporarily, or so he thought, he abandoned his crystals to help her. She was pleased and flattered and grateful.

The problem before them now was to separate the new element from pitchblende. Since pitchblende had been carefully analyzed by other scientists and its composition was known, Marie knew there must be only a very small quantity of the mysterious substance to have escaped their attention. Perhaps, she told Pierre, there would be no more than one percent of it present. This seemed to her at the time a modest estimate. In her wildest imagining she did not guess that the proportion would turn out to be not even one millionth of one percent.

For the separation process, they developed a new method of procedure, based on radioactivity. First, they separated the compounds of the pitchblende ore by the usual chemical analysis. Then they measured each of the compounds for radioactivity. Those that were not radioactive they could discard. Gradually the radioactivity became more concentrated in the substances they kept and continued to work with. They could not take away their new element from pitchblende, for they didn't know what to look for. They had, step by step, to take the pitchblende away from the element. Very slowly their search narrowed down.

After a while it became evident that the radioactivity was concentrated both in ores containing bismuth and in those containing barium. This was unexpected, too, and led them to another startling fact. They were actually on the track of two new elements instead of one.

In July, three months after the first notice to the Academy of Science, the Curies were able to announce the existence of one of these elements, the one that had an affinity for bismuth. Marie named it *polonium*—after her native land.

It becomes very hot in Paris in July. In their crowded office-workroom—they could not dignify it by calling it a laboratory—the heat was insufferable. They had earned their vacation and were off to the Auvergne again, this time taking with them ten-month-old Irène. It was pure joy to be able to be with their little girl when they were not exhausted with fatigue, and they were as proud of the fact that she had cut her seventh tooth and could stand half a minute alone as they were of the discovery of polonium.

In September they returned to Paris and their workshop and plunged themselves with renewed enthusiasm into their experiments. Finally, in December of 1898, shortly after Marie's thirty-first birthday, they were able to announce in the Academy of Science *Proceedings* the discovery of the second new element, to which they gave the name of *radium*.

"The new radioactive substance certainly contains a very strong proportion of barium," they wrote. "In spite of that, its radioactivity is considerable. The radioactivity of radium therefore must be enormous."

CHAPTER SIX

# A Wooden Shack

Though Pierre and Marie no longer had any doubt as to the reality of polonium and its sister element, radium, many of their fellow scientists were skeptical. Elements must be seen and studied under a microscope if they are to be believed. They must be given an atomic weight.

No one had seen radium or polonium. Were they light, like helium, with its atomic weight of 4.003? Or heavy, like uranium, which has an atomic weight of 238.07? Who could believe that an element which is made up of matter, whether it is a gas, a liquid, or a mineral, could have the quality of radiation, the more opinionated members of the scientific profession demanded. Why, next someone would be claiming that the light rays of the sun were formed of matter. The Curies were upsetting all the preconceived scientific beliefs.

In 1898 Marie and Pierre didn't have the slightest idea of the atomic weights of radium and polonium, nor of what their elements looked like in pure form. Neither could they explain the reason for their peculiar quality of radiation. Both of them were bound and determined they would find out.

With all the work they had already done, they were still in the predicament of looking for something about which they knew nothing, except for the feature of radioactivity. As before, their experiments had to proceed by the trial-and-error method.

So far they had been working with a fairly small supply of minerals. Now that they were beginning to see in what minute quantities their new elements existed, they realized they would probably need tons of ore, preferably pitchblende. Pitchblende was mined extensively at St. Joachimsthal in Bohemia, Austria, where uranium salts were extracted for use in the manufacture of glass. In its natural state, this ore was very costly. They could never afford to buy all they needed.

Marie came up with a solution. Why not try to purchase the residue of pitchblende after the uranium salts had been extracted? The residue had no commercial value, and it would still contain radium. Through the Academy of Science at Vienna, they sent their request to the Austrian Government, who owned the mines.

The Austrians acted promptly and with surprisingly little red tape. They would be glad to give a ton of pitchblende residue to the two French scientists. All the Curies needed to do was to pay for the cost of transportation. If they wanted more later on, they could arrange to get it from the mines at a minimum charge.

Marie never forgot her debt of gratitude to the Austrian Government.

Their second concern was the always present one of a place to work. They had outgrown their office and had to have more space. They needed, in fact, a modern laboratory, evenly heated, with airtight doors and windows to keep out dust and dirt. It didn't seem too much to ask, considering the value of their work to science and to France.

But all their efforts to get the use of a satisfactory laboratory from government or university sources were as unsuccessful as before. Again it was Director Schulzenberger of Pierre's school who came to their rescue.

Across the courtyard from the workroom where they had their electronic installation was an old, abandoned wooden shack. It had a floor of soft asphalt and a glass roof that leaked. The furnishings consisted of some worn pine tables, a cast iron stove that smoked badly—and a blackboard. Previously it had been used as a dissecting room for medical students, but had long since been found not even suitable for that. The only good thing about this shack was that no one else wanted it. It was all theirs.

Soon after they moved in, a heavy wagon drove up with their ton of pitchblende from Bohemia. Pierre and Marie rushed out, bareheaded and in their laboratory smocks. Pierre, as usual, was calm, but Marie could not control her feelings.

"*Our* pitchblende, Pierre!" she cried.

As the first sack was unloaded, she cut the strings and plunged her hands in to pull out the drab-colored rocks that contained their wonderful element.

"What color do you think it will be, Pierre?"

He didn't need to ask what she meant by "it." "I think it will have a beautiful color," he told her, smiling.

The next day they set to work in earnest. Again, they had to separate the pitchblende into all its separate compounds and examine each for its radioactivity, but on a much larger scale than in their early experiments. They used all known methods of chemical analysis—boiling, filtering, addition of acids—and they invented new methods of their own.

For the first two years in their makeshift laboratory, they worked together on every step they took. This included the physical labor of handling the ore, as well as the study of the radiation of the increasingly active products they were obtaining. In a series of little notebooks covered with black cloth, they made notes of each of their chemical treatments and analyses. Sometimes these records were in Marie's clear, neat writing, sometimes in Pierre's almost illegible scrawl. The notebooks, still in existence, give little indication either of the days and weeks when everything went wrong, or of their successes.

There were no outlets to carry away the poisonous gases thrown off in the chemical treatments. When the weather permitted, they could conduct some of their experiments out in the court. When it stormed, they did them inside, leaving doors and windows wide open. In winter, near the stove, the heat was stifling. A yard away from it they would be freezing. In summer, with the fire from the stove, they almost suffocated.

Thus two years passed, and the end of the project they had taken on with such enthusiasm was still not in sight. The first ton of pitchblende was gone, and they had ordered several more from the mines at St. Joachimsthal at their own expense. The substances they had separated were remarkably radioactive, but the spectrums showed that they were nowhere near having radium in its pure form. As for polonium, they had abandoned for the time being any attempt to isolate it. It existed in even smaller quantities than radium.

Still their search had not been altogether unfruitful, as several scientific papers they published in this period proved. One was a memorandum on the discovery of the induced radioactivity produced by radium. Another was on the effects of the rays. A third dealt with the electric charge carried by certain of the rays. They had also made a general report on new radioactive substances and their radiations for the Congress of Physics, which met in Paris in 1900, and in addition, Pierre published a study of the action of a magnetic field on radium rays.

They loaned samples of their radioactive products to Henri Becquerel, the discoverer of uranium rays, who was now their close friend, and to other scientists. One such sample reached Adam Paulsen's Expedition in Northern Iceland, and Paulsen wrote them a warm letter of thanks, saying that they were now using the Curies' "radiant power" to conduct their meteorological tests. French scientists, as well as foreign ones, were following their work with the greatest of excitement, and some were conducting their own experiments on radioactivity.

With all this, the Curie family was still trying to live on Pierre's 500 francs a month, which he received for his lectures at the School of Physics and Chemistry. Though old Dr. Curie continued to look after Irène during their long hours away from home, they had to keep a servant to do the housekeeping. It was an impossible situation.

The very qualities that Marie loved in Pierre, his modesty and his integrity, were a handicap when it came to the problem of increasing his earnings. He didn't know how to ask for favors. The Chair of Physical Chemistry at the Sorbonne fell vacant in 1898, and Pierre had dutifully applied for it. He was turned down. Since he was neither a graduate of the Normal School nor of the Polytechnic, he did not have the backing these big schools gave to their pupils. Later he was granted the position of tutor at Polytechnic, but this post lasted only six months. Somehow they struggled on.

Then in 1900 came a fabulous offer. The University of Geneva wanted Pierre to take a chair in physics there at a salary of ten thousand francs a year. In addition, he would have an allowance for their living quarters. He would have the direction of a fine laboratory and two assistants. Marie, too, would be given a position in the same laboratory. The University wrote they would consider themselves fortunate to have among their faculty a man and a woman in the first rank of European scientists.

It struck them as ironic that their first formal recognition should be from a country other than France. The fact remained that the proposition was very tempting indeed. That summer Pierre and Marie went to Switzerland to visit the university and make up their minds what to do. They were received with the warmest of welcomes and spent several delightful weeks looking over the laboratory that could be theirs and meeting their Swiss colleagues. After all that, they decided to return to Paris. To accept the university's offer would mean to postpone their search for radium for an indefinite period, and that they could not do.

When they got back, they found that the mathematician Henri Poincaré, who admired Pierre and wanted him to stay in France, had used his influence to get him a Chair of Physics at the P.C.N. This was the name given to a series of special courses in physics, chemistry, and natural history for medical students, held in an annex of the Sorbonne on the Rue Cuvier.

About the same time, Marie, who had received her teaching certificate several years before, was finally assigned a position teaching physics at the Normal School for Girls at Sèvres, outside of Paris, near Versailles.

Temporarily their financial problem was solved, but it was at a great cost in time and effort. Pierre had two outside jobs now instead of one; Marie had to take the long tram ride to Sèvres several times a week. Her night work now included correcting papers and preparing lectures, in addition to charts and tables on radium. They had worked themselves to a point of exhaustion before, and they did so now, with the added gnawing regret that they simply could not spend as much time in their laboratory as they felt they must. They were always tired, and it was worse for Pierre, who about this period began to suffer with pains in his back and legs so severe he often could not sleep at night.

Another chair at the Sorbonne fell vacant, this one in mineralogy. This was something for which Pierre was more than qualified because of his publications on the theories of the physics of crystals. Had he been accepted, he would have been able to quit his post both at the School of Physics and at the P.C.N. and work fewer hours at more money. And he would have had some sort of decent laboratory. But once again his application was turned down. It seemed that whatever he and Marie accomplished, they must do it the hard way.

Beginning with this third year of their search for radium,

they decided to divide their efforts in an attempt to make faster progress. Pierre continued with the study of the properties of radium, which was the part of their work that came under the classification of physics, while Marie carried out the chemistry part, of trying to isolate radium salts. This meant that her share was largely physical labor.

Sometimes she worked with forty pounds or so of material at one time. The shack would be filled with huge vessels of precipitates and liquids. It was more than exhausting to lift the containers and transfer the liquids, not to mention the long hours in which she stirred the boiling substances with an iron bar.

She fell ill with pneumonia and for a couple of months couldn't get out of bed. As soon as she was able, she was back at the laboratory. It tortured Pierre to see the struggle she was going through, and at one point he suggested they abandon what had begun to seem a thankless task. Marie would not listen to him. They had started something, and they were going to finish it.

Finally the time came when she had extracted from all their vast mounds of ores the barium-bearing radium. This, in a state of chloride, she submitted to the process of purification through fractional crystallizations.

The crystallization process consisted of evaporating a solution of radium and barium chloride. When a certain concentration was surpassed, crystals formed on the surface. More radium was present in the crystal phase than in the liquid phase. By repeatedly dissolving the crystal phase and recrystallizing it, the radium content became higher. It was a long and tedious process, but they could find no quicker method. More than fifty years later, modern scientists had not perceptibly improved the Curies' method, except that bromides were used instead of chlorides.

These last stages were less strenuous physically. They could keep the concentrated substances in small vessels, and it was no longer necessary to carry great cauldrons out into the courtyard.

But it was now that they felt most keenly the inadequacies of their shack. The work was precise. More than ever they needed an even temperature and an insulated laboratory if all was to go smoothly. When dust or particles of iron and coal fell into the products purified with such painful care and the experiments had to be repeated, Marie gritted her teeth. It seemed like the final straw. Or rather, like one last challenge to her stubborn determination.

With all their trials, these were happy days for them both. Once they entered their shack, they seemed to be in a world of their own, remote from the stir and bustle of the city. Often they ate a simple lunch of bread and cheese as they worked. When they were cold, a cup of hot tea, drunk beside the stove, revived them. While supervising some of their many operations, they would walk back and forth, talking of their work and what it would mean to the future of science. Never had they been so close together.

Most of the time they were alone, though occasionally they got outside assistance. There was a laboratory helper named Petit, a simple man who was devoted to them both. He helped them with the carting away of the refuse and other chores. A laboratory technician in the School of Physics named G. Belmont worked with them from time to time. Then there was the young chemist, André Debierne, whom Pierre encouraged to study radioactivity on his own. He did so and found another new radioactive element, *actinium*. Actinium was never a very important element, but its discovery paved the way for scientists to look for other radioactive substances. Pierre also collaborated with Georges Sagnac, a young X-ray physicist, on the investigation of the electric charge carried by secondary rays.

Occasionally other of Pierre's colleagues and students would stop in, and there would be discussions, usually led by Pierre standing at the blackboard. In his classrooms as in his laboratory, Pierre liked to have a blackboard before him so he could illustrate the matter under discussion. In later years quite a number of prominent French scientists traced their early zeal for scientific research to those sessions with Pierre. The tall, gaunt figure, bent slightly forward but not stooped, his supple gestures, the white chalk with which he wrote down pertinent formulae as he made the detailed explanations of some difficult problem—all were vivid memories that these men retained through the years.

Though few people at this time knew the name of Marie Curie, Pierre's reputation was spreading. In 1901 the Academy of Sciences awarded him the Lacaze Prize, a considerable honor. Then early in 1902, a fellow scientist proposed him as a member of the elite Academy of Sciences.

According to tradition, each candidate for membership had to pay calls on those who were already Academy members and tell each in turn why he was qualified. Pierre felt that this was a ridiculous procedure. It would be much more sensible—and more dignified—if the Academy would elect new members without forcing them to make such calls.

He agreed to go through with the distasteful visits only when his friends pointed out that Academy membership would make it much easier for him to get a post as a Sorbonne professor. But when he presented himself to the Academy members, he found it quite impossible to speak of his own achievements. Instead, he talked at length about the abilities of his opponent. In the end, the opponent received thirty-two votes, whereas Pierre got only twenty.

Marie was afraid he would be disappointed, but his reaction was simply disgust that he had wasted so much valuable time.

Outside of their colleagues, Marie and Pierre saw practically nobody in this period of their life. Even their occasional pleasant evenings with Bronya and Casimir were denied them. The Dluskis had now returned to Austrian Poland, where at Zakopane in the Carpathian Mountains they were building up a sanatorium for tuberculosis patients. There were no other social activities that tempted them in the slightest as their excitement mounted in the last lap of their long search.

The only time they didn't resent spending away from the shack were the hours with little Irène at home. And frequently, after Marie told stories to her daughter until she fell asleep, they would by mutual consent return to the haven of their laboratory. In these latter months, before they turned on the lamp, they could see a luminous glow coming from the various vessels arranged on the tables and shelves along the wall.

One fall evening of 1902, nearly four years after they had first moved to the shack, they looked at each other triumphantly across the table. The tons of pitchblende that had passed through their hands in various stages, in cauldrons and iron basins and smaller vessels, had finally yielded up the hidden element. It was all there now—a tenth of a gram of chloride of radium, hardly enough to fill the tip of a teaspoon (one gram equals about one-eighteenth of an ounce).

Already they had made an estimate of its atomic weight as 225 (later established as 225.97), less than the atomic weight of uranium, but still among the heaviest of the elements. Pierre had said he thought it would have a beautiful color. It was not colored at all, it turned out. In the broad daylight, it looked very much like common table salt. But in the darkness, as they had seen in these last stages, it was transformed, giving off a strange iridescent light. There was nothing else like it in the world.

Here, finally, was the proof for which the more conservative scientists had been waiting. Radium did exist.

## CHAPTER SEVEN

# *Fame*

ON JUNE 25, 1903, SOME five years after Marie had first decided on the subject of her doctor's thesis, she made her formal application for her degree to the board of examiners, in the students' hall of the Sorbonne. The paper on which she was being judged was called "Researches on Radioactive Substances," and was signed *Madame Sklodovska Curie*.

The examiners were Marie's former professor, Gabriel Lippmann, himself a physicist and the developer of color photography; the heavily bearded Henri Moissan, who invented the electric arc; and Professor Bonty, another brilliant scientist. Other prominent physicists and chemists were in the audience of the small amphitheater.

All eyes were fixed on Marie. A woman applying for a degree of doctor of science was an unusual sight under any circumstances. But that a young and attractive woman should have assembled such a mass of original information was truly amazing.

Her hair was still as fair and soft as when she had first come to Paris, though her face was slenderer and more mature. A few lines etched in her smooth high brow marked the strain of the past four years. Even more telling of her labors were her hands, chapped like those of a cleaning woman from her years of bottle washing and with curious burns on her fingers that never seemed completely to heal.

To Pierre and old Dr. Curie and to Bronya (who had come from Poland for this occasion) she had never seemed more beautiful. Bronya in particular watched her with an almost motherly pride. For once she had put her foot down and made Marie buy a new dress, a becoming one of black wool and silk. Marie couldn't understand the need for it. The old one, shabby as it was, had seemed quite adequate to her. The dressmaker's fittings had made her bored and impatient.

It was a pity Professor Sklodovska could not witness the triumph of his youngest daughter. He had died the year before, having spent his last years with his son, Joseph, looking after his grandchildren.

There was nothing bored or impatient about her manner now. Nor would any person present, seeing this serious young woman for the first time, have suspected how she had run out into the street to plunge her hands deep into a sack of pitchblende, or how her eyes had shone with delight at the luminous glow of "her" radium.

Calmly and precisely she answered the questions of her judges, sometimes turning to write down an equation on the blackboard. She described the radioactive properties of polonium and uranium. She talked about radium.

The whole atmosphere of the hall was one of deep solemnity. No one smiled. No one showed either amazement or approval. It would have been contrary to tradition. Yet no one doubted that Marie Curie's paper was the greatest single contribution to the history of science that had ever been made in a doctor's thesis.

Finally the questioning was over.

Professor Lippmann, the president of the board, gave the verdict in the usual formula: "The University of Paris grants to you the title of doctor of physical science, with 'very honorable' mention." He added in his own words, "In the name of the jury, madame, I wish to pass on to you the congratulations of all of us."

The session was over. People filed out quietly as was proper. But once outside, they gathered around Marie—the members of her own family, the scientists, and even a group of schoolgirls who had come from Sèvres to witness their sweet and gentle professor receive this high honor. How proud they all were of her! As for Marie, she felt only relief that the public ordeal was over and she could return to the tranquility of her laboratory.

In spite of Pierre's bad health and the strain of his two teaching jobs, in spite of Marie's own fatigue and the classes at Sèvres and her household duties, they did not stop their research at this point. In all, they published 32 scientific reports on their investigations.

The news of Marie Curie's radium was discussed everywhere in scientific circles.

This strange new element was about five million times more radioactive than its equal weight in uranium, though the proportion of radium in uranium ores was hardly more than three-tenths of a gram of radium to the ton of uranium.

The salts of radium were self-luminous. They shone in the dark like tiny fireflies.

Another incredible thing about the element was that it spontaneously gave off heat. In an hour it produced enough heat to melt its own weight in ice. Or, to put it in another way, the heat given off by radium was 250,000 times as much as that produced by the burning of an equal weight of coal.

It could make an impression on photographic plates through black paper. It turned the atmosphere into a conductor of electricity. It colored the glass containers that held it in shades of violet. When it was wrapped in paper or cotton wool, it corroded these substances until they were reduced to powder. Its rays could penetrate glass, cloth, wood, and even harder substances.

It could give phosphorescence to other mineral products. A diamond, for instance, became luminous if exposed to the rays of radium.

It also transferred its radioactivity to objects near it. This was a most curious phenomenon. In Marie's laboratory, the spread of radioactivity had to be constantly checked, for it interfered with the accuracy of the delicate precision instruments. Radioactivity affected the air and dust in the room, the clothes of those who worked in the laboratory, practically everything.

This new element also had harmful qualities. A tube containing a grain the size of a pinhead paralyzed a mouse in three hours when it was placed over the creature's spinal column. In fifteen hours it died.

Henri Becquerel once made the mistake of carrying a few radium salts in a glass tube inside his vest pocket. It gave him a nasty burn. "I love your radium, but I have a grudge against it," he told Marie.

As an experiment Pierre had exposed his arm to the rays of radium, and a sore had appeared that had taken a long while to heal. Marie suspected too that the burns on her fingers were the result of her daily contact with radium. As yet, neither of them guessed the risk they had run in their daily work with radium and radioactive substances.

They did reason that if radium could affect healthy tissue, it could be used to destroy unhealthy tissue. With two medical scientists, Pierre conducted experiments on animals, which gave evidence that radium, properly used, might cure growths, tumors, and even cancer. This possibility pleased Marie more than anything else.

The work of the Curies inspired scientists everywhere to follow the same line of research.

Some of the most brilliant work in the new science of radioactivity was being done in the Cavendish Laboratories in England, where a group of brilliant young men were working under J.J. Thompson, the discoverer of electrons, or negative particles of electricity.

One of Thompson's disciples was a young man from New Zealand named Ernest Rutherford. Rutherford, reading of the discoveries of Roentgen and Henri Becquerel and the Curies, decided this would be his field of study. By 1900 he had found that the radioactive element *thorium* gave off a gas that was itself richly radioactive. The name he gave to this gas was "emanation," a name that Madame Curie first used for radon, a similar gas that came from radium.

Rutherford left the Cavendish Laboratories to go to McGill University in Canada, and there with Frederick Soddy, a scientist with a mind as keen as his own, he published in 1902 a new theory of radioactivity. The exciting thesis of this report was that atoms of radioactive elements were not stable. They were constantly changing and withering away, and as they did so, they threw off particles of matter. The young men did not yet know what these particles were, but they recognized that the constant explosion of atoms was beyond man's control. Neither the extreme cold of liquid air nor the intense heat of an electric furnace influenced the disintegration any more than did chemical treatment.

In the case of radium, the disintegration took place at the fantastic rate of 35 billion atoms per second in each gram! Even so, it takes 1,620 years to reduce itself to half; that is, the half-life of radium is 1,620 years, as compared with uranium, which has a half-life of four billion some years. The disintegration of uranium forms radium.

Here indeed was a brand-new conception of matter— as something living and changing and giving birth to new substances, and destroying itself.

Scientists at this time were split as to the cause of radium's rays. Lord Kelvin, Pierre's old friend, suggested that the radioactive atom absorbed unknown radiations from all of space. But Rutherford, Thompson, and the young French scientist Jean Perrin held to the theory that the energy of radioactivity came from within the atom, as was later generally accepted.

Letters came to the Curies' laboratory from the greatest scientists in Europe—from England, Germany, Austria, and Denmark—asking for information, and Marie and Pierre considered it their duty to answer these requests in fullest detail. The manufacture of radium was also developing. In 1904 a French industrialist named Armet de Lisle started a radium factory to furnish doctors with the precious substance, enlisting the services of Marie and Pierre for this undertaking.

Even money and proper equipment did not make it easy to wrest radium from its hiding place, and in the first few years after its discovery, the cost went up to $150,000 for a single gram.

One day Pierre received a letter from a group of American technicians in Buffalo, New York, asking for information about the method of processing radium. This was purely for commercial purposes and brought up a new problem.

Pierre discussed the matter with his wife. "If we wish, we can consider ourselves the 'inventors' of radium. We can patent our techniques so we can share the profits wherever radium is produced."

She stared at him. "It is impossible to do such a thing. It would be contrary to the spirit of science."

Pierre agreed with her completely. Much as they needed money to continue their research and get their own laboratory, they could not do it this way. He had felt obliged to let Marie make the final decision about this, but once she had done so, he wrote immediately to the Buffalo technicians, telling them everything they wished to know. His letter to them was shown to Marie many years later when she paid her first visit to America.

Though in this way they gave up all chance of making a fortune from their discovery, renown was gradually making an inroad into their lives. In 1903 Pierre was proposed for the Legion of Honor, an award that he refused, saying that he felt no need to be decorated—all he wanted was a laboratory.

Later that same year he and Marie went to London at the invitation of the Royal Institution. Lord Kelvin, a man of youthful enthusiasm in spite of his advanced years, was there to welcome them and to introduce them to other noted scientists, including William Crookes and William Ramsay. Pierre gave a lecture on radium at the Royal Institution, and though Marie wasn't asked to speak, she was allowed to attend the session—the first time the institute had ever admitted a woman.

A few months after that, Pierre made a return journey to London, where he accepted the Davy Medal, the highest award of the Royal Society of London, in Marie's name as well as his own. There was no place in their home for the impressive medal. In the end they gave it to little Irène as a plaything. She was delighted with her "big gold penny."

A month later, on December 10, 1903, they received another honor, one of much more practical value than any medal. The Academy of Science at Stockholm announced that Marie and Pierre Curie, together with Henri Becquerel, were to receive the Nobel Prize in Physics for their discoveries in radioactivity.

The Curies' share of the Nobel Prize would be around seventy thousand gold francs, which it was "not against the spirit of science" to accept.

Their first expenditure was to hire a laboratory assistant so they would not have to wait for help promised by the university. There were other pleasurable things to do with this small fortune. Marie sent 20,000 Austrian crowns to Bronya and Casimir as a loan, to help with the work of the sanatorium. They gave presents and loans to Pierre's brother, Jacques; to Marie's sister Hela; to needy Polish students; to one of Marie's young girls at Sèvres;

and to others. For the house on Boulevard Kellermann they installed a modern bathroom and repapered one of the rooms. They put something aside for the furniture.

But for herself, Marie did not even buy a new hat.

## CHAPTER EIGHT

# *Catastrophe*

MARIE AND PIERRE WOULD HAVE enjoyed receiving the Nobel Prize tremendously if the award could have been slipped to them quietly, without anyone knowing about it. The trouble was that the giving of the Prize was worldwide news, even as it is today. Overnight Marie and Pierre became famous.

Their story made unusually good copy. One reason was that Marie was the first woman to receive the Nobel Prize. The fact that she and Pierre had worked together and that she was also a homemaker and a mother added interest. They had found it a nuisance and a handicap to carry on their experiments in a miserable leaky laboratory, but to their public there was something romantic about their four long years of drudgery.

They were besieged with reporters and photographers from every country in the world. The press wasn't content with knowing what radium and radioactivity were—this they could have learned by reading scientific reports. In fact, it sometimes seemed that was what interested people least of all. The world wanted to know what the Curies looked like, what they wore, what they ate—all the details of their personal lives that editors call "human interest material." Representatives of the press even took pictures of little Irène's black-and-white cat.

The laboratory became a place of pilgrimage for the curious. Fashionable society women came, sniffing disdainfully at the falling plaster and holding up their skirts to protect themselves from the mud in the alleyway. Newspapermen wrote of the "narrow, dark, and deserted street," the "twisted tree dying in a corner of the courtyard," and "the deep, melancholy silence that pervaded the enclosure." In fact, their descriptions made the peaceful laboratory sound like an excellent setting for a crime scene.

Both at their home and the laboratory, mail flooded in like a tidal wave. Most of it was from autograph hunters, unsuccessful inventors, those who demanded money or wanted the Curies to back some pet scheme. But all of it had to be read and sorted, and the letters from scientists and those who had legitimate requests had to be answered.

Everyone in Paris wanted to entertain them, and there were a few invitations they couldn't refuse. One evening they were summoned by Monsieur Loubet, the President of the French Republic, to dine at the Élysée Palace. Royalty from all over Europe were present, and the story is told that during the evening a lady came up to Marie and asked if she would like to meet the King of Greece.

"I don't see any point to it," she said, quite honestly.

The woman looked at her aghast, and suddenly Marie realized she was speaking to her hostess, Madame Loubet. When the wife of the President of the Republic asks one to meet a king, one does not refuse. "Of course, I shall do whatever you wish," Marie stammered in embarrassment.

The truth was that Marie and Pierre had neither taste nor time for public life. The laurels heaped on their heads didn't flatter them. At a reception once, when Marie was wearing her only evening gown, a very modest one compared to the elaborate dresses of the other women, Pierre looked at her admiringly.

"It is a pity," he said. "Evening dress becomes you." Then he added, shaking his head, "But of course there is no time."

Time was the most valuable possession they had. Only with time could they carry on the work that was more important to them than anything else. After they became famous, everyone seemed to feel they had the right to the time of the Curies. In distress Pierre wrote to a friend that the very persons who were asking him to write articles and make lectures would be astonished when years had passed and they had done no work.

Privacy became a thing of the past. With all their hardships they had been happy in their solitude at the laboratory. Their evenings at home with Dr. Curie and their little daughter had been relaxing and satisfying. That was the sort of life they were meant for. Now all that had vanished.

They took the only escape that seemed possible. With their bicycles they set off on a trip to Brittany, hunting out tiny villages where they were not known. At night they stayed in country inns under false names. Happily, most people are convinced that celebrities wear fine clothes and travel in style. No one suspected that the tall man in rough country clothes and the youngish woman in peasant dress were the "sages of the Rue Lhomond."

An American journalist who was on their trail caught up with them at Pouldu, in Finistère. Marie was sitting on the stone steps of the fisherman's cottage where they were staying. She was barefoot, shaking the sand from her slippers. The journalist took her for a country woman until she looked up, and he recognized her from her photograph.

He began to plague her with questions. Patiently Marie resigned herself to the inevitable. Yes, she was Madame Curie, she admitted. She and her husband had discovered radium. As long as he kept to questions of fact, she answered him briefly but courteously.

But the man had his heart set on getting an exclusive interview of the sort that would make his American editor sit up and take notice. He began to ask her about her childhood, about how it felt to be a woman scientist, and other personal questions.

She stood up. "In science one must interest oneself in things, monsieur, not in people," she said firmly and fled into the cottage.

One good thing came to them from the Nobel Prize, in addition to the money. France finally recognized Pierre's abilities officially. A new chair was created for him at the Sorbonne, and at the beginning of the school year 1904–05, Pierre Curie was named Professor of the Faculty of Sciences.

This was only half of the triumph it should have been, for while he had been voted a professorship, nothing had been done to get him a laboratory. Even the mediocre one at the P.C.N. was better than nothing. Pierre wrote to the heads of the Sorbonne that he had decided to remain where he was. This gesture stirred them to action, and a fund was voted for a laboratory and laboratory helpers. The fact that the fund was woefully small was offset by the fact that Marie was named chief of his laboratory. In giving this lowly position, which paid just 2,400 francs a year, to a Nobel Prize winner and the discoverer of radium, the officials of the Sorbonne acted as though they were doing her a favor.

As a professor at the Sorbonne, Pierre was finally able to give up his post at the School of Physics and Chemistry. His replacement there was Paul Langevin, his former student and an excellent scientist in his own right.

On December 6, 1904, the Curies' second child was born, a little girl whom they named Ève Denise. Bit by bit their life began to swing back to normal. By one way or another, they shut themselves away from overzealous reporters and had some peace and quiet.

When little Ève was six months old, her mother and father

took a long-postponed visit to Stockholm. The terms of the Nobel Prize required that the recipients give one lecture there, but because of fatigue and illness they had been unable to fulfill this task before.

Before the Stockholm Academy of Science, Pierre spoke of the consequences of the discovery of radium. In the study of rocks, it gave the key to mysteries that had never been explained. Because of radium the established principles of physics had to be modified. In chemistry it opened a wide field of speculation about the source of radioactivity's energy. In medicine, it was proving useful in the treatment of cancer.

It was in this lecture that he foresaw that radium might lead to the discovery of a terrible type of explosive, "a fearful means of destruction in the hands of great criminals who lead people toward war." He further stated, "I am among those who think, with Nobel, that humanity will obtain more good than evil from the new discoveries."

Both Marie and Pierre enjoyed this trip to Sweden. The contact with Swedish scientists was stimulating. They fell in love with the lakes and pine forests and little houses of redwood, and they came back refreshed.

In this same year, Pierre was persuaded to offer himself as a candidate for the French Academy of Science for the second time. He hated calling on the members as much as he had before, nor was he any more capable of bragging about himself. His reluctant campaign succeeded, and at long last he was voted a member of the Academy. It was by a disgracefully small majority. There were still a number of Americans who considered Pierre Curie, who lacked the conventional educational background, not qualified to join their ranks.

He ignored the slight and plunged himself into his work at the Sorbonne. For this he was allowed to choose his own subjects. He gave some lessons in radioactivity, and he gave other lectures on the subject he had neglected since he had joined

Marie in her work on radium—the laws of symmetry and the application of these laws to the physics of crystals. It was his aim to work out a course that would completely cover the physics of crystallized matter, a subject little known in France. At the same time, he was engaged in a collaboration with A. Laborde on investigations on mineral waters and gases discharged from springs. This was his last published work.

Marie had never been more in love with her husband and more proud of him than during this period. It seemed to her that his intellectual faculties were at their height. His natural curiosity pushed him to undertakings in different directions, and he changed the object of his research with surprising ease. Often he did not bother to publish the results of such research. That this sometimes resulted in others getting credit for work he had done before them didn't worry him in the slightest. In this he was the pure scientist, for whom jealousy was a stranger.

He was fascinated by his two children and never tired of trying to understand these little beings. As Irène grew older, he took her for walks, treated her as an adult, and took great delight in watching the development of her young mind. Perhaps he already guessed that she would be able to carry on his and Marie's work.

He and Marie had few intimate friends, and they were all scientists. On Sunday afternoons they often gathered in the Curie garden—Paul Langevin, Georges Sagnac, André Debierne, Madame and Monsieur Jean Perrin. The Perrins lived next door; their gardens were separated by a fence covered with rose vines. Madame Perrin was Marie's closest woman friend. Jean Perrin was a physicist, well known for his studies of cathode rays. Their children and the Curie children were inseparable.

In Easter of 1906, the Curie family had a wonderful holiday in Brittany. As when they were first married, they took their bicycles into the woods, returning with great bouquets of flowers. Other days they stretched out on the grass, watching with delight as fourteen-month-old Ève tottered after Irène, who was dressed in boys' clothes and chasing butterflies with a green net.

Marie and the two little girls stayed on a couple of days after Pierre returned to Paris. The day they got back, he and Marie attended a dinner of the Physics Society, where he had a long conversation with Henri Poincaré, discussing, among other things, methods of teaching. It was Pierre's belief that teaching should be based on experience and contact with nature. Poincaré, referring to that conversation later, wrote: *I admired the fecundity and the depth of his thought, the new aspect which physical phenomena took on when looked at through that original and lucid mind.*

The next day, April 19, 1906, it was raining. Marie stayed home with the children, while Pierre attended a reunion of the Association of Professors of the Faculty of Sciences. One of the things the members talked about had to do with accidents in the laboratories. Pierre suggested a plan to make the work safer for researchers.

It was still raining when he left his friends, and he carried a big umbrella as protection against the downpour. Undoubtedly he was not giving much attention to where he was going, his mind on some problems connected with his work.

As he turned, suddenly, to cross the street, he found himself directly in the path of a heavy wagon drawn by two powerful horses, coming from Pont Neuf. He slipped on the wet streets, fell, and while the horses' hooves missed him, the wheel of the wagon did not. His death was instantaneous.

Passersby stared in horror. In one fleeting instant, the great mind and the great person that had been Pierre Curie was lost forever to the world.

Paul Appell, now Dean of the Sorbonne Faculty, came with Jean Perrin to bring the sad news. Marie was out, and old Dr. Curie was alone in the house with the servant. He took one look at their faces and guessed the truth. Silently the three men sat and waited for Marie's return.

# CATASTROPHE

At six o'clock they heard her key in the lock. She entered, fresh and blooming, with cries of joy at seeing her old friends there. Then she stopped short, sensing that something was wrong.

As gently as they could, they told her what had happened.

She stood motionless. "Pierre is dead?" she asked finally. She waited before them accusingly, as though to force them to deny a fact that was too terrible to accept.

## CHAPTER NINE

# The Long Struggle

If Marie had been able to cry her heart out, it might have eased her sorrow. Instead, the news of Pierre's death seemed to turn her to stone. At the funeral ceremonies in Sceaux, she stood by Pierre's tomb staring blankly ahead of her like an automaton. Someone placed a bouquet at the foot of the tomb. Suddenly she seized it, and one by one tore off the flowers and scattered them over his coffin.

How could she go on without him? It seemed impossible to her at first. All her married life she had been surrounded by his unique tenderness. It was too cruel abruptly to lose that tenderness.

She remembered that she and Pierre had once talked about the possibility that one of them would die before the other. Marie had announced flatly that she could not exist without him. He had looked at her gravely and had said, "Whatever happens, if one should become like a body without a soul, still one must always work."

She knew she had to continue because he would have wanted her to. In her secret diary, she wrote long letters to him, as though he were still alive. Somewhere she had to find the strength and the courage not only to live but to work.

She first told young Irène that her papa had been hurt badly in the head and had gone away for a rest. A few days later she let her know the truth. Irène cried a great deal but then seemed to get over her grief. Ève, the baby, was still too young to understand at all.

Right after the funeral, the French government officially proposed to grant Marie and her daughters a pension. Marie refused. "I am young enough to earn a living for myself and my children," she said.

Letters and telegrams of sympathy poured in from fellow scientists, from kings and heads of state, and from common people. The whole world mourned his loss.

The heads of the university had their own problem. Who would succeed Pierre Curie? What man was capable of continuing the courses that he had started so brilliantly? Paul Appell and a few others argued that there was no man capable of carrying on where Pierre had left off. The only person who could do it was Marie Curie, his widow.

On May 13, 1906, the council of the Faculty of Science unanimously decided to give to Marie the Chair in Physics created for Pierre. She would receive a salary of 10,000 francs a year, sufficient to support her daughters comfortably. This was the first time in the history of France that a post in higher education had been offered to a woman.

Marie thought it over. At first it seemed to her that she could not face such an ordeal. Finally she decided she would try, writing to Pierre of her decision in her diary.

The house on Boulevard Kellermann, filled with memories of her husband, became unbearable, and she moved to Sceaux. Pierre's father went with her. To the end of his days, he stayed with this Polish woman whom his younger son had married and whom he had come to love as his own daughter.

Marie's first lecture at the Sorbonne took place at one thirty on November 5, 1906, a little more than six months after Pierre's death. It was announced in a newspaper that the subject of her lecture would be an explanation of the theory of ions in gases, and that the lecture would also treat of radioactivity. The same paper complained that this historic lecture would be given in the regular lecture hall instead of in the big amphitheater of the Sorbonne. The lecture hall had only seating space for 125, and there would be almost that many students. Surely seats for the press and the public should be provided, the article concluded.

By noon of that day, the lecture hall was filled to capacity and the overflow spread out into the corridors and the courtyard. Never in the history of the Sorbonne had any lecture attracted such an audience. The new President of the French Republic, Armand Fallières, was there with his wife. The English scientists, Lord Kelvin, William Ramsay, and Sir Oliver Lodge, were present, as were King Carlos and Queen Amelia of Portugal. There were other statesmen, titled ladies, members of the Academy, and men and women from all walks of life.

Of course many who came did so not because they were interested in "the theory of ions in gases," but simply from curiosity. They wanted to see how Madame Curie would conduct her lecture. Would she thank the Sorbonne for allowing her to hold such a position? Would she begin by praising her predecessor, as was the custom of the university? But the curiosity seekers were disappointed.

At the appointed hour, she slipped through the back door and onto the platform—a slender, pale woman with large eyes, dressed in black. The brilliant audience rose, and a roar of applause burst forth. She held up her hand in an appeal for silence. Then, fingering the little gold chain around her neck, she began in her low voice, with hardly a trace of Polish accent:

"When we consider the progress that has been made in physics in the past ten years, one is surprised at the advance that has taken place in our ideas concerning electricity and matter..."

Her listeners gasped. Marie Curie had taken up Pierre's lecture at the exact point where he had left off.

In addition to her courses at the Sorbonne, Marie continued to teach her girls at Sèvres. Nor could she neglect her duties as a mother. Now that their father was gone, her children needed her guidance and care more than ever.

She wanted them first to grow up strong and healthy, and in her new home she had a trapeze installed in the garden. Both girls became expert trapeze artists. They learned swimming in their vacations on the Brittany coast. Several times Marie sent them to visit Bronya and Casimir in the Carpathians, where they developed skills in horseback riding, mountain climbing, and skiing.

Irène early showed a talent for mathematics. Ève's first interest was music. Marie noticed these trends with pleasure. Like Pierre, she had definite theories of education. She intended for her girls to learn a great deal, but she didn't want them to spend long hours in stuffy classrooms as she had done. She talked over her ideas with her colleagues, and they devised a plan.

One day a week Irène and several sons and daughters of Marie's friends trooped off to a Sorbonne laboratory where Jean Perrin gave them an elementary lesson in chemistry. The next day they studied mathematics under Paul Langevin. They took classes in sculpture, drawing, literature, and history under men of similar reputation. On Thursday afternoons, in a small room at the School of Physics and Chemistry, these youngsters had the rare privilege of attending a class in physics under Marie Curie.

It was a challenge to her to keep her explanations simple so that basic principles of physics could be grasped by these young minds. This she did by having them conduct their own experiments. For instance, they placed bicycle ball bearings, dipped in ink, on a sloping surface, to illustrate the principle of a parabola. Another time she had them make a thermometer, under her supervision.

Several of her child pupils, Irène among them, at a later date decided their future lay in physics, perhaps partly due to the inspiration of working with Marie. The novel experiment lasted just two years before it was abandoned. For the children it was a marvelous thing.

At the same time, the number of Marie's students at the Sorbonne increased steadily. The wealthy American Andrew Carnegie gave several annual scholarships for her work in 1907, which provided her with additional helpers in her laboratory and new young minds to train, including her nephew, the son of Jacques Curie.

Serious students were fascinated by her lectures. She would bring up every argument for and against a certain position, using both theory and experimental evidence. When she finished there would not be a single link missing in her case, and she would be ready to draw her conclusions.

Some of the male professors, like actors, put on a good show in their classrooms, dramatizing themselves and their subject. Marie did not have this gift. She spoke in a monotone, rather fast, and it required her students' full concentration to keep up with her.

They all respected her and at the same time felt protective toward her. Once in a while, a reporter would find his way into the Sorbonne halls, in search of an interview with the world's most famous woman scientist. The students, knowing how she feared and detested such interruptions, would gang up against him. One of them would go ahead to warn Marie so she could go in hiding. The others got rid of the intruder.

With all this activity at home and at school, Marie managed to find time to continue her own work. In 1908 she collected and edited a 600-page volume, *The Works of Pierre Curie*, a labor of love.

About the same time, she undertook, with the help of André

Debierne, an experiment that everyone said could not possibly be done. This was to isolate the actual radium metal. So far only radium salts had been isolated. No one had yet attempted to find the metal. She tried numerous methods of separation, all unsuccessful. For weeks and months she practically lived in her laboratory. That she had only a few tenths of a gram of radium chloride to work with made her task even more difficult.

One day in 1910, she passed an electric current through molten radium chloride. At the negative mercury electrode, she noticed a chemical change. An amalgam or alloy was being formed. She heated this alloy in a silica tube filled with nitrogen under reduced pressure. The mercury boiled off as a vapor, and then before her eyes at last was the elusive radium metal—in the form of brilliant white globules that tarnished in the air!

This was an extremely delicate and complicated operation, as there was constant danger of losing the radium if it wasn't done with the utmost care. Nor did she keep the radium metal in this form long but returned it to its chloride state, as it was needed for research purposes. Only once since that time has this experiment been repeated. For Marie, it was the crowning achievement of her scientific work.

Polonium, her first new element, continued to interest her, and in that same year of 1910, she and André Debierne finally isolated two milligrams of it to the state where it was nearly five percent pure—a remarkable feat of skill and persistence. It was discovered that polonium, far more radioactive than radium, had a half-life of only 140 days, as compared with radium's 1,620 years. Actinium, the radioactive element discovered by Debierne, had a half-life of some twenty years, much shorter than they had previously thought.

She went on to other problems—and found a way of measuring the quantity of radium by the emanation gas that it produced. In this manner she was thus able to measure less than a thousandth of a milligram with fair precision. This was of the greatest importance since the success of radium treatments for cancer depends on the exact quantity of radium being used. A commission of scientists of different countries was formed to set up an internal standard. Marie was appointed to prepare the first standard sample, which was accepted by the commission and deposited with the Internal Bureau of Weights and Measures at Sèvres.

Several other such samples were later put into service by the commission. One was in the charge of Marie's laboratory, where anyone could bring radium to be tested. In the United States, this would be handled by the Bureau of Standards.

About this time Marie was proposed for the Legion of Honor, but turned the honor down as Pierre had done. She was also proposed for the Academy of Science, and like Pierre, was persuaded to make personal visits to Academy members. But as had happened to Pierre on his first try, she lost out by a few votes. The Academy was not ready to admit women to its membership. The experience left her with a feeling of distaste, and she resolved she would never go through it again under any circumstances.

Five years had passed since Pierre's death, and in these five years Marie had succeeded in disproving what the jealous ones had hinted—that the discovery of radium was Pierre's and that she was not capable of carrying on creative research by herself.

Once again recognition came to her from a foreign country. Sweden granted her another Nobel Prize, this time in Chemistry, for the isolating of radium metal. It was the first time that the Nobel Prize had been awarded twice to the same person. Bronya and Marie's daughter Irène went with Marie to Sweden to receive the award. Though her health was far from good, she attended the innumerable affairs in her honor, including the King's dinner and a reception by the women of Sweden, and her Swedish hosts were enchanted with her modesty and unassuming manners.

On her return she found she no longer had the strength to

commute daily to Sceaux. She moved her small family back to Paris, to an apartment at 36 Quai de Bethune, on the Ile St. Louis. This little island, back of the Cathedral of Notre Dame on the Seine, is one of the quietest and most picturesque parts of all Paris, but Marie had not been there long when she fell very sick indeed and had to be taken to the hospital for an operation.

Her recovery was long and slow; for weeks she was so feeble she could hardly stand. Then in May of 1912, something happened that helped her more than any of the rest of the cures the doctors ordered.

A delegation of Poles came to see her, headed by Henryk Sienkiewicz, author of the novel *Quo Vadis* and a Nobel Prize winner like herself. The delegation told her that a new building for the study of radioactivity was being built in Warsaw and that they wanted her to come and be their director.

She did not accept, but she did go to Warsaw for the opening ceremonies. The Russian control over Poland was less severe than in her youth, though the Russian officials pointedly ignored her arrival. Their coolness was in contrast to the wild welcome given her by the Polish people.

The next summer, in 1913, she was well enough to take a walking trip in Switzerland along with some friends and her two little girls. One of the party was her fellow scientist, Albert Einstein. Many years later Ève described tagging along behind the two of them on the rocky trail while they carried on an excited conversation about theoretical physics.

All of a sudden Dr. Einstein seized Marie by the arm and stopped short, saying, "You understand, what I need to know is exactly what happens to the passengers in an elevator when it falls into emptiness."

To Irène and Ève, his remark sounded so ridiculous they burst into howls of laughter. He was dead serious. The imaginary fall in the elevator posed problems directly connected with his "theory of relativity."

The French people often say of themselves that it takes them a long time to be goaded into action but that when they get started, nothing will stop them. This proved true in the case of their treatment of Marie. For years they had treated her more or less like a problem stepchild who must not be ignored but could not be completely accepted. Finally things started happening.

For several years plans had been underway to build an Institute of Radium in the university grounds on the newly renamed Rue de Pierre Curie. All sorts of red tape were involved; it seemed it would never be completed. Then Dr. Pierre-Emile Roux, Director of the Pasteur Institute, offered to build a laboratory for Marie. She would thus have been working for the Pasteur Institute instead of the Sorbonne.

Oddly enough, this gesture enraged the Sorbonne authorities. Marie must be kept with them at all costs, they stated. A compromise was eventually reached between the two groups. The Sorbonne and the Pasteur Institute would both contribute to two buildings—a laboratory of radioactivity, to be called the Radium Institute and to be directed by Marie, and a laboratory for biological research and Curietherapy (as treatment by radium was sometimes called) under the physician Dr. Claude Regaud.

As though to make up for the previous neglect of her, Marie was allowed to have her say about the construction of the new buildings. She was pleased, though she couldn't repress a pang of bitterness that they had waited until after Pierre's death to provide the laboratory that would have meant so much to him.

Her plans for the new Institute would have suited her husband. She wanted it to be a place not only where she could do good work herself but that would serve for future young scientists. The research rooms must be large, with wide windows to let the sunlight stream through. She insisted on an elevator. For the garden between the two buildings, she ordered lime and plane trees. She planted rambler roses there herself, watering them every morning.

One day the laboratory helper, Petit, who had been so good to her and Pierre at the School of Physics and Chemistry, came to tell her that the old school was being remodeled and that they were tearing down her shed. She went back with him to have a final look. It was the same as when she had seen it last, even to the blackboard that still carried some formulae in Pierre's handwriting. Momentarily the cruel reality of his death seemed a bad dream. She half expected his tall figure to appear and to hear the sound of his familiar voice.

Shortly after that the splendid *Institut du Radium, Pavillon Curie* was completed to everybody's satisfaction. Everything was ready to move in by July 1914—just before the outbreak of the First World War.

CHAPTER TEN

# *War*

BY SEPTEMBER PARISIANS, PARTICULARLY THOSE with money to travel, were leaving their city in droves. They had heard of the horror of the siege of Paris when the Germans had come in 1870—some of the old-timers had lived through it. The German army seemed just as mighty under Kaiser Wilhelm II as it had under Bismarck. Already they had invaded little Belgium and made a victorious march through the valley of Oise toward Paris. The French government had departed to Bordeaux.

To live in Paris as it had been in 1870—without lights, without theaters or amusements, with the food supply running out—many shuddered at such a thought. The national road south was jammed with motor cars. The trains were equally crowded.

A passenger on one of the trains was Marie Curie, sitting quietly in a corner on a wooden bench. This fragile middle-aged woman with workworn hands might have been any overworked housewife. She carried with her only a small overnight bag and a case that looked lighter than it was. It was made of lead and was made to order for its contents—a small vial enclosing a gram of radium. The government had charged her with taking it to Bordeaux for safety.

This port on the Atlantic coast was a pleasant town in peacetime. Now it was crowded and noisy. Taxi cabs were seized by the most aggressive. Hotel rooms were impossible to find. Night was approaching.

A government employee who had taken Marie under his wing offered to try and find a room in a private home for her. She waited for him in the public square outside the station, beside the valuable case that it was beyond her strength to carry. Would she have to stay there all night long, she wondered, as the crowds jostled and pushed her. It was, she decided, an amusing predicament.

Her friend finally returned with the news that he had a place for her to stay the night. The next morning, after she had put her radium into safe hands, she took the train back for Paris.

At the station people stared at her, not because she was Marie Curie, which they didn't know, but simply because she was a civilian going to Paris instead of fleeing from it. Several persons spoke to her and seemed relieved and comforted that she found it perfectly natural to return.

The train moved slowly, with many delays. For several hours it was stranded on the rails. It was a troop train filled with newly mobilized soldiers. One of them, who must have seen that she looked faint, offered her a piece of bread. She devoured it almost greedily. Only then did she remember that she had eaten nothing since the day before when she left her laboratory.

Paris had never looked so lovely to her as when she left the station after her fearfully exhausting journey. There was good news too. The German advance had been turned back at the Battle of the Marne. Paris, then, was not going to fall, like a ripe fruit, into German hands, as the frightened had expected.

Ève and Irène had passed the summer in Brittany, where

Marie had originally planned to join them. Now she felt safe in letting them come home and continue their studies.

At this time the government was urging everyone to do his duty, but in the case of civilians, even scientists, no specific instructions were given as to what that duty consisted of. Marie soon found her own niche.

It was known by this time that X-rays offered doctors and surgeons an invaluable means of examining the wounded and finding the exact location where bullets or shell fragments had lodged. Yet the Military Board of Health had no department of radiology, and radiologic installations existed only in a small number of major hospitals.

Beginning with the outbreak of war, Marie made a check of university laboratories and manufacturers to find out just how much radiological equipment there was. Then, by persuasion and argument, she got it transferred to various hospitals and enlisted volunteer helpers whom she instructed in how to operate it. Even by the time of the Battle of the Marne, these few new radiological stations were able to save a great many lives.

With the help of the Red Cross, Marie fitted up an ordinary touring car to carry a complete radiologic apparatus. A dynamo, attached by a long cable to the apparatus, was worked by the car engine to furnish the electric current necessary to produce X-rays. The car was sent to hospitals without such equipment, where it was needed.

This was the beginning of one of the greatest and least publicized single contributions to the war effort. It also marked the temporary transformation of Marie Curie from a shy and retiring person into a woman who was afraid of nothing and of nobody.

After she returned from Bordeaux, she set to work to get more automobiles that she could transform into mobile radiological stations. In some cases she persuaded wealthy women to lend her their limousines for the duration, promising she would return them after the war—provided they were still in working order. She demanded, and got, contributions to purchase other vehicles.

In all, she succeeded in equipping twenty motorcars for her service. Limousines and less impressive makes alike, they were all painted a standard gray, with a Red Cross and a French flag painted on their plates. The soldiers nicknamed them "little Curies."

One of them, a Renault with a body like a truck, she kept for her own use. She had a military chauffeur assigned to her, but because he was not always available, she took lessons in driving. She learned how to change a tire and how to clean a dirty carburetor. At the same time she was perfecting her own knowledge of X-rays and giving herself a course in anatomy, for to give lessons to doctors in the X-raying of the wounded, she had to know as well as they did the structure of the human body.

The gray Renault became a familiar sight on roads leading to the front, and some of the sentries, but not all of them, would recognize the woman next to the chauffeur, or at the wheel, as Madame Curie.

With varying details, each of her trips was like the next.

A call would come in that a new load of casualties had arrived at a certain field hospital. The chauffeur loaded up with gasoline while Marie checked the dynamo and other equipment to make sure all was in working order. Then they were off. Their journey might be all day or all night. They were sure to be stopped many times and asked for their papers.

When they reached their destination, there was no time to rest, for human lives were at stake. Marie quickly decided on the room most suitable as a radiological hall, hung up black curtains or an army blanket at the windows to darken it, and returned to the car to unpack her instruments while the chauffeur unrolled the cable connecting the apparatus with the dynamo. He sat at the wheel and, at Marie's signal, started the motor while she tested the current and prepared the radioscopic screen. It usually took half an hour, more or less, to get everything ready.

With the surgeon at her side, she waited as, one by one, stretchers were brought in and the wounded men placed on the table. Sometimes they would be unconscious. Sometimes they would be moaning with pain. Marie steeled herself not to collapse at the sight of the blood and the terrible wounds.

Some of the patients were frightened. X-rays, they had heard, made fearful burns. Marie reassured them with gentle and consoling words. Their X-ray would be no more painful than an ordinary photograph. She did not tell them, if she knew, that she was in more danger from the rays than they were. For while there is no risk at all in one short exposure, repeated exposure to X-rays can produce very grave injuries.

Some of the surgeons, unfamiliar with X-rays themselves, were astounded to see the plates reveal the bony structure underneath the flesh, and Marie would have to point out the dark spot that was the bullet or shell fragment. In drastic cases a surgeon might operate at once, guided by the picture on the radioscopic screen. Or Marie sketched a copy of the plate or made a quick photograph of it for the surgeon's use.

Hour after hour passed in this fashion. The procession of stretchers seemed endless. "To hate the very idea of war, it ought to be sufficient to see once what I have seen so many times," Marie wrote later.

After the last stretcher had been taken away, Marie talked to the hospital chief, promising them X-ray equipment of their own. By one way or another, she always managed to keep such promises. Many times she loaded the apparatus on the train herself with the help of railway workers, to make sure it would go forward immediately. Often she returned personally to the hospital to install the equipment and to train the manipulators how to use it.

In all, during the war she supervised installations of X-ray equipment to 200 hospitals. Thus, with her twenty "little Curies," France could claim 220 radiological posts, stationary and mobile, due to Marie's efforts. In the course of the four years at war, more than a million men had X-rays made of their wounds.

During this period Marie was hardly ever at home. She visited ambulance stations of the armies of the north, at Amiens, Calais, Dunkirk, and Boulogne. She went to Verdun, Nancy, and Compiègne. She was sent into Belgium and once worked at the Hospital in Hoogstade with Queen Elisabeth and King Albert of Belgium, whose simplicity and devotion to the care of the sick impressed her greatly. In contrast were certain society women doing voluntary work in hospitals near Paris for the thrill of it; they would take one look at Marie's unstylish clothes and decide she was a person of no social position. On several of her trips, Marie took along her older daughter, Irène, who at seventeen was beginning her studies at the Sorbonne and was also studying nursing and radiology.

The journeys never went smoothly. There was always uncertainty of finding lodging and food, or of even being allowed to proceed. Valuable time was wasted seeing military commanders to obtain passes and permissions for transport. In compensation were the friendships built up with men and women who, like themselves, were giving their services without counting the cost.

One of the most difficult problems was to find trained assistants. The X-ray equipment could be handled by any intelligent person who knew how to study and had some notion of electrical machinery. Professors, engineers, and university students were at various times commandeered into the service. The trouble was that these men were often transferred by military orders after they had been trained, and the search had to be started again.

Beginning in 1915 Marie had begun to move her laboratory equipment to the new Institute of Radium. She had to do it little by little, sometimes using her radiological cars for transportation. Afterward, with the help of Irène and her mechanic, she classified and arranged her materials in the new working quarters.

When everything was in order a year later, she gave her first course there. It was done with the cooperation of the Nurses' School of the Edith Cavell Hospital, and very fittingly, it was on radiology. She and Irène trained some 150 operators, all women, who could be sent out to the radiological posts Marie had established, and who would not be called away for other military duty.

Marie set into operation another project. Her gram of radium had been brought back from Bordeaux in 1915. She wanted it to be used to treat the sick and wounded without, however, risking the loss of the precious substance. Thus she placed at the disposal of the Health Service not the radium, but the emanation of the radium, preserved in tubes. It was used for treating scars and skin injuries and in many cases proved more practicable than the direct use of radium.

When the course for the women X-ray operators was completed, Marie allowed Irène to do ambulance service at the front. The young girl worked at Amiens, where the German cannons caused the beautiful old cathedral to tremble with shock, and later at Ypres, where the enemy made their first use of poison gas against French soldiers. At the end of the war, Irène was given a medal for her services.

After the failure of the German offensive in the summer of 1918, Marie, at the request of the Italian government, went to Italy to study their natural resources in radioactive material. She stayed a month and succeeded in interesting Italian public authorities in the importance of this subject. When she returned to Paris, she found she had twenty new students for her radiological classes—young soldiers of the American Expeditionary Forces.

She was in the institute laboratory on November 11, 1918, when the guns sounded the signal of the armistice. Promptly she left everything and rushed out to buy some French flags. There were none to be had. She managed to purchase some red, white, and blue cloth, which her cleaning woman sewed together and hung at the windows. Then Marie and two of her assistants climbed into her old Renault and drove through the streets of a rejoicing Paris. At the Place de la Concorde the crowds were so thick the car hardly moved. Cheering strangers climbed up on the top of it and rode with them. On that day in Paris, there were no rich, no poor, no famous, no failures. Everyone was united.

Marie had two reasons for great happiness. One, of course, was that the long parade of stretchers that brought wounded men to her X-rays was ended. The other was that with the end of the war, Poland, her country even after all these years, was free again. Polish children would no longer have to attend Russian schools, nor would Polish patriots be punished for their loyalty to their country. Those "born in servitude and chained since birth," as Adam Mickiewicz, the Polish poet, had written, could cast off their chains.

All this was reward for the sacrifices the war had demanded of her—the loss of her savings, the strain on her health, and worst of all the four-year neglect of her own research.

She was fifty years old and tired and sick and nearly penniless. But the war had not deprived her either of the brilliance of her mind or her stubbornness. With those assets she started over again.

## CHAPTER ELEVEN

# *America*

ONE OF THE BEST-KNOWN WOMEN in America in the 1920s was Mrs. William Brown Meloney, who edited the women's magazine, *Delineator*. She was tiny and old and quite frail, and she limped slightly from a childhood accident, but she commanded respect and attention and sometimes fear.

Powerful as she was in her own circle, she was humble before those she considered truly great. In her editorial box was a slip of paper on which she had written, *Greatest Woman's Story in the World—Marie Curie, Discoverer of Radium.*

From 1915 on she sent her most promising writers to Europe with instructions to get a story about the French scientist. All of them brought back excuses: Madame Curie was away at the front; her bags of mail lay unanswered at the institute; personal publicity was distasteful to her.

Finally Mrs. Meloney went to Paris herself to try and get an interview. She wrote ten letters and tore them up. The one she finally sent was very short. *My father, who was a medical man, used to say that it was impossible to exaggerate the unimportance of people,* she wrote. *But you have been important to me for twenty years, and I want to see you a few minutes.*

The next morning she was admitted to Marie's office at the Institute of Radium.

Mrs. Meloney had her own ideas of how famous scientists lived. She had visited Thomas Edison's magnificent laboratory, and she had been a neighbor of Alexander Graham Bell and admired his splendid house and fine horses. She knew, too, what a tremendous industry radium had become. In Pittsburgh she had seen the smokestacks of the greatest radium reduction plant in the world. Millions of dollars were spent on radium watches and radium gun sights, and millions of dollars' worth of radium was stored in the United States.

It seemed logical that the woman responsible for all this should live in a palace, and it was a shock to find the tiny office into which she was ushered as bare and plain as a nun's cell. The door opened and Marie entered. "A pale, timid little woman in a black cotton dress, with the saddest face I had ever looked upon," the American wrote later. "Her kind, patient, beautiful face had the detached expression of a scholar. Suddenly I felt like an intruder."

Mrs. Meloney, who had interviewed celebrities all her life, for once found herself at a loss for words. It was Marie who, seeing her embarrassment, started the conversation.

There were some fifty grams of radium in America, she said. Four of them were in Baltimore, six in Denver, seven in New York. She went on naming the location of all the rest.

"And in France?" Mrs. Meloney asked her.

"My laboratory has one gram," Marie told her.

"You have only one gram?"

"I? Oh, I have none. This gram belongs to my laboratory."

Impulsively Mrs. Meloney asked her, "If you had the whole world to choose from, what would you take?"

The answer was obvious. The only thing Marie wished for was

an additional gram of radium for research purposes. But it was an impossible wish. Radium was much too expensive.

Mrs. Meloney returned to America with a marvelous plan, to get the women of America to donate a gram of radium to Madame Curie. It cost $100,000 at that time, and her first thought was to find ten wealthy women who would give $10,000 each. A major donation came from Mrs. William Vaughn Moody, widow of the famous playwright and poet, but other women of wealth were not so generous. Undaunted, Mrs. Meloney formed the Marie Curie Radium Fund, which campaigned for funds from all American women, rich and poor. The full amount was soon collected.

Then, out in the southern part of Colorado toward the end of 1920, a gang of 300 men working under their foreman, Joseph M. Flannery, set to digging tons of yellow ore called *carnotite*. Carnotite, like pitchblende, contained radium. Five hundred tons of this ore were hauled by wagon and burros 18 miles to a concentration mill, which had the nearest water supply. Here the ore was chemically treated until only 100 tons were left. These were crushed into powder, packed into hundred-pound sacks, and shipped 65 miles to Placerville, where they were loaded into railroad cars to travel 2,500 miles to Canonsburg, Pennsylvania.

At the Canonsburg plant, 200 more men, trained in the handling of chemicals, reduced these 100 tons to a few hundred pounds. All that was left was then sent under guard to the Standard Chemical Company in Pittsburgh, where there remained the final task of isolation. The result of this massive effort was just one gram of radium salt.

Five hundred men with the best equipment available, following a proven process, had worked a year for that one gram—just ten times as much as Marie and Pierre had isolated by themselves in their damp and rotting shed on the Rue Lhomond, learning each step as they went! It was because of this prodigious labor that radium became the most precious substance in the world—a hundred thousand times more valuable than gold.

Mrs. Meloney wrote Marie that her gram of radium was ready for her. Would she consider coming to America for it? The idea appealed to Marie, but it also terrified her. She replied that her health was not good and that, moreover, she did not like to leave her two daughters.

She should bring Ève and Irène with her, Mrs. Meloney wrote back promptly. They would arrange a program for Marie that would tire her as little as possible. So that they would not be bothered by the irksome details of traveling, Mrs. Meloney herself went to Paris to cross back to America with them.

The two daughters were thrilled, and Ève actually got her mother to buy a couple of new dresses. They set sail on the *Olympic* in the spring of 1921, and they were given the most luxurious suite on the ship. Mrs. Meloney had seen to that.

But their friend was not able to prevent the enormous crowd gathered at the pier to greet the visitor. It seemed as though everyone in New York had turned out. A delegation of three hundred women waving red and white roses represented Polish organizations in the United States. There were contingents of Girl Scouts and schoolgirls. And there were, of course, scores of reporters, photographers, and cameramen. Ève and Irène had never before quite realized what a world personage their quiet mother was.

At home Marie had developed certain techniques for avoiding journalists. Here she was helpless. They gave her an armchair on the deck of the *Olympic* where she could at least sit down while the photographers focused their camera on her and journalists hurled questions she was too startled and dazed to answer.

Somehow Mrs. Meloney managed to get her away and up to the quiet of her own apartment. It was overflowing with roses. A grateful horticulturist, who had been cured of cancer by radium, had been growing them especially for her.

The next day, she had luncheon in Mrs. Andrew Carnegie's beautiful home. That was a "must" on the agenda, for Marie had not forgotten what she owed to the American philanthropist, who had died two years before, for the scholarships he had given her students.

An itinerary was drawn up for her. Every city and every college and university in the United States was eager to claim her. Medals and honorary doctor degrees and other titles were awaiting her by the dozen.

She would need her cap and gown, Mrs. Meloney reminded her. They were essential for the ceremonies. Marie confessed she didn't own such garments. The men of the Sorbonne wore gowns, but since she was the only woman, no precedent had ever been set for her. A tailor was called in immediately, who fashioned an impressive gown of black silk with velvet facings. Marie found it hot and uncomfortable. The sleeves were a nuisance, and the silk irritated her fingers, scarred by radium. Though she put it on when the occasion demanded it, she never would wear the headdress that went with it.

Her first jaunt out of New York proved unexpectedly pleasant. She was taken to the eastern women's colleges, Smith, Vassar, Bryn Mawr, and Mount Holyoke, where thousands of young girls lined the grassy slopes of the campuses, waving flags and carrying flowers.

This first glimpse of American college youth pleased her. It seemed a fine idea to have colleges in the country, where young people could go swimming and boating and play tennis and baseball. She was struck with the physical education courses that were a part of American college curriculum. She liked the smiling eager faces of her young hostesses, and she recognized the emphasis on initiative in American college life.

More formal appearances were awaiting her back in New York. At a huge assembly at Carnegie Hall, American scholars and the French and Polish ambassadors paid her tribute, while college delegates presented her with American Beauty roses and lilies. One of the guests was an old friend, Ignace Paderewski, whom she remembered from the time when she was a poor student and he a struggling pianist. In the next few days there was a lunch of chemists at the Museum of Natural History, a banquet at the Waldorf-Astoria, and innumerable other affairs. Already she was exhausted, and her American visit had scarcely begun.

All that had happened so far was a prelude to the big day when President Harding, at a ceremony at the White House, would present Marie with her radium.

The evening before this ceremony, Mrs. Meloney showed her the parchment scroll that would accompany the gift. Marie read it carefully and then said there was something there that must be changed. The scroll said that the gift was for her personally. It must read that the radium was a gift to her laboratory.

"Of course," Mrs. Meloney murmured. "It is a mere formality. We will take care of it next week."

Marie insisted the change must be made at once. Suppose she died within a few hours. The radium would then be left to private persons in custody for her daughters. That was out of the question. The radium must be used for science.

With some difficulty, a lawyer was found at that late hour, and the wording of the parchment was changed.

Thus it was that the gram of radium from the women of America became not a gift to Marie Curie but to her laboratory. For that matter, the lead-lined casket on display in the White House East Room on that day of May 20, 1922, did not even contain the radium. Even within its lead casing, radium was considered too dangerous a substance for non-scientists, particularly the President of the United States, to handle. It was the scientist Robert Andrews Millikan, inventor of the electrical balance, who presented the actual gram of radium to Marie, at the National Museum of Washington at a later date.

But these were details that did not interfere with the impressiveness of the White House ceremony. Diplomats, Supreme Court judges, top officers of the Army and Navy, presidents of universities, and other notables attended. After innumerable speeches, President Harding passed to Marie the roll of parchment tied with a tricolor ribbon and slipped over her head a silken cord on which hung a golden key—the key to the casket.

Other ordeals, some interesting, some merely trying, awaited her. Before she left Washington, she was invited to attend the opening of the new Laboratory of Mines, created to do research on very low temperatures by use of liquid hydrogen and helium. She visited several big radiotherapy hospitals and noted with envy their fine laboratories for the extraction of radium emanation.

At Pittsburgh she visited the radium factory where the final processes were conducted for extraction of radium salts. This was interesting but not too different from the way it was done in her laboratory in Paris. She was scheduled to receive an honorary degree at the University of Pittsburgh but was too sick and tired to go the short distance to Carnegie Institute where the ceremony was to be held. Irène went as her proxy and accepted the parchment in her place.

Later she was driven 25 miles to Canonsburg, to the mill where the carnotite ore was brought for initial processing. When she looked over the vast plant and realized that the same work was being done here as she and Pierre had done in their small shack, she was more than overwhelmed. Though she could not go a short distance a little while before to receive a university degree, she spent three solid hours walking through every department of the plant. The American technicians were astounded at this triumph of mind over physical weakness.

The next day she could not get out of bed. Doctors insisted she abandon the scheduled western tour. Journalists wrote passionate editorials about how Americans were killing their noted guest with hospitality. But in a few days she had recovered sufficiently to take a trip to Niagara Falls and then on to Buffalo. She visited several other cities. She escaped waiting crowds by leaving the train at the rear end and stepping over the rails. Irène and Ève received degrees for her. It struck sixteen-year-old Ève as very funny when distinguished orators, addressing speeches to her intended for her mother, referred to her "magnificent work" and her "long life of toil."

Though their formal western tour was called off, Mrs. Meloney took them to the Grand Canyon. There Ève and Irène rode wiry little ponies along the crest of the chasm, exclaiming with delight at the ever-changing coloring, from violet to red and from orange to pale ocher.

In Chicago, Marie was made an honorary member of the University of Chicago. In New York, Columbia University gave her the title of doctor *honoris causa*. She found reserve strength to visit Harvard, Yale, Wellesley, and Radcliffe, where more honors were heaped on her.

On June 28, after only a little more than a month, she was allowed to leave. Her cabin in the *Olympic* was filled with flowers and telegrams. The ship carried her gram of radium too.

Of all her memories of her first American trip, there were three she guarded most tenderly: the views of Niagara Falls and the Grand Canyon, and one small reception she had attended in Chicago. It was held in the Polish quarter for Polish émigrés. For them Marie was a symbol of their fatherland, and men and women gathered around her to try to kiss her hands or touch her dress.

CHAPTER TWELVE

# Scientists Unite

Usually Marie kept away from politics. Scientists should devote themselves to science; politics were for politicians, in her opinion. The war changed her thinking somewhat. When politicians created war, they were interfering with the free advance of science.

She followed the work of the League of Nations with great interest. It was obvious that the League was not perfect, but it was a first step in the right direction. It represented a hope for the future.

In May of 1922, the Council of the League of Nations asked her to join the newly formed International Committee on Intellectual Cooperation. The aim of this Committee was to "facilitate intellectual exchange between nations, particularly in regard to communication of scientific information." Which meant, more simply, that it would give scientists of all nations a chance to get together and exchange ideas.

Marie hesitated. By nature she was not a joiner, but this seemed important to her. In her indecision she wrote to her old friend Albert Einstein. He not only urged her to join the

Committee but told her he was joining himself. Like Marie, Einstein had a strong sense of social duty. He believed that by working with the Committee, they could accomplish some good for science and for humanity.

The chairman of the Committee was the French philosopher Henri Bergson, and its membership included the best minds of the world. The meetings, held in Geneva, Switzerland, discussed things like tying together astronomic measures in different countries, setting up standards for telegraphy, and creating an international meteorological office. One of Einstein's pet projects was an international university where, in particular, history could be taught without national prejudice.

The project most dear to Marie's heart was setting up scholarships for young scientists, to be financed by governments and private donors. In June of 1926, she presented her complete program for this undertaking.

In this period Europe was in a state of turmoil. The heavy reparations demanded at the Treaty of Versailles had forced Germany into poverty and a sense of inferiority that was laying the groundwork for Hitler's war machine. In Italy there was fascism and Mussolini. Unfortunately, this atmosphere of uncertainty was reflected within the Committee, devoted as it was supposed to be to pure scientific exchange.

In 1925 the League replaced an Italian anti-fascist member of the Committee with Rocco, the Italian fascist Minister of Justice. The more liberal members of the Committee were enraged at this political appointment. Marie opposed it on the grounds that a government official did not belong with the scientists. Einstein felt that Rocco should be kept out since he represented a government where liberty of opinion was strangled and where intellectuals were persecuted.

This was a sample of the type of argument that intruded on the meetings. Both Einstein and Marie soon felt that too much time was wasted on matters that had nothing to do with science. Once she wrote to him, rather sadly, that she felt that he was wrong in believing that people with cultivated minds could communicate with each other more easily than others. On the contrary, it seemed to her that common men were able to exchange ideas more freely than intellectuals.

One night, after a particularly stormy session of the Committee, Einstein and Marie strolled down to Lake Geneva and sat on a bench. A streetlamp made a silvery streak on the rippling water.

"Why does the reflection break on the water at this spot and not at another one?" Einstein asked suddenly.

His question was enough to start them both off into a discussion of mathematical equations and laws of physics. As they talked, they became relaxed and peaceful. They were in their own world of science now, where all was harmony and reason.

Over the years Einstein and Marie saw each other only infrequently. They were both very busy people with many responsibilities. Even when they were in different parts of the world, they exchanged letters from time to time. Marie continued her interest in her plan to have students subsidized so they could work without money worries, and once she asked Einstein's aid to find an organization to give a grant to a brilliant young Jewish scientist who worked in her laboratory. In their letters they talked about world affairs, about mutual friends, hardly ever about their work or themselves. Nonetheless, these letters give concrete evidence of their high regard for each other.

As a child and as a young girl, Marie had been warm and friendly. In her years of study and work, she had had to repress her social instincts, and after the death of her husband, she had shut herself off completely from the world for a long time.

Her war work had forced her out among people, and following the war, she gradually allowed herself a more normal sort of life. She always remained shy, but her natural kindliness and sympathy became more evident. She never stopped sorrowing for Pierre, but time, as it always does, mellowed that sorrow so that it stopped being an acute pain.

Watching her daughters grow up was a rich experience for Marie. They were very different. Ève was the prettier one. She loved nice clothes and parties, for which Irène did not care a whit. Ève was musical and literary; she had hordes of friends outside of scientific circles. From her seventeenth year, and perhaps even before, Irène never swerved from her ambition to be a scientist.

Ève's liveliness and Irène's dependable good sense made a good contrast for the small family of three, still living on the Quai de Bethune. A startling change came in the pattern of their lives when, in 1926, Irène told her mother she was getting married. It was unexpected. Irène had never seemed the slightest interested in young men—except for a dark-haired young student named Frédéric Joliot, who had come to work in her mother's laboratory a couple of years before.

Like Pierre Curie, Frédéric Joliot had attended the School of Physics and Chemistry. Paul Langevin, who had replaced Pierre there, had recommended him to Marie, and when he reported to her for work, he was still in uniform, just completing his military service. He hadn't known as much about radioactivity as Irène, and she had undertaken to teach him herself.

He had confided to her that he had venerated Pierre and Marie Curie ever since he had been a small boy. A small engraving of them that he had kept in his room had been his inspiration. Many of the laboratory helpers found Irène cold and distant, but Frédéric saw that beneath her reserve she had a poetic nature and a great genius. He began walking home with her evenings after work, and sometimes they took Sunday excursions out into the country. One day he proposed to her.

After they were married, the big apartment seemed empty to Ève and her mother. Marie could console herself that she had found a son. It was a delight to see him and Irène doing original research, though she refused to favor these two over her other assistants.

As Director of the Institute of Radium, Marie's task now was not so much her own research as supervising the research done by her youthful students and helpers. They would be waiting for her, young men and women in white laboratory blouses, when her chauffeur brought her to work in the mornings. They always had questions to ask or experiments they wanted her to see. Marie loved it. If sometimes she passed along her own ideas and let her students work on them, it was the least of her concern. It was what the institute as a whole could accomplish that counted. Under her influence the institute published 483 scientific papers between 1919 and 1934. Of these, thirty-one were credited to Marie, though she undoubtedly had a hand in many others.

Though her health remained uniformly poor, often she was away on trips. Her voyage to America had taught her that her presence and the use of her name were valuable in getting new hospitals and research laboratories under way. She could no longer deny herself to the world.

She went to Italy and Holland and England and Spain. Once she stayed with President Masaryk at his country house in Czechoslovakia and was pleased to learn that his tastes were as simple as her own. In Belgium she dined informally with King Albert and Queen Elisabeth, her friends from World War One. Her most adventurous trip was to Rio de Janeiro, the beautiful seacoast capital of Brazil, where she spent four weeks with Irène.

She made several more trips to Poland to visit Bronya, a widow now like herself, her brother Joseph, a prominent physician, and Hela, happily married and a teacher. Marie's dream was to establish a radium institute in Poland similar to hers in Paris. But Poland was poor after its long period of enslavement. It seemed impossible to raise the funds.

Bronya, though no longer young, organized a nationwide campaign. BUY A BRICK FOR THE MARIE SKLODOVS-

## SCIENTISTS UNITE

KA-CURIE INSTITUTE read the posters and postcards that flooded the country. Brick by brick the institute grew, and finally the building became a reality. But it still lacked one essential—a stock of radium.

Marie remembered her American friend, Mrs. Meloney, and wrote her about her problem. Once again Mrs. Meloney called on the women of the United States, and the money, just $50,000 this time, for the price had been lowered, was soon found. This gram was not taken from the carnotite of Colorado. Belgium had found a rich supply of pitchblende in the Congo, and a draft for the money was prepared for Marie so she could buy her gram from that country. But once more Mrs. Meloney persuaded her to come to America to receive her gift.

She set sail for "Dollaria," as her friend Einstein called the land of American dollars, in October 1929.

Her welcome was as warm as on her first visit, though she was physically unable to repeat her earlier tour. She stayed several days at the White House, where President Hoover was now President, and then, escorted by Mrs. Meloney and the financier Owen D. Young, she went by a private railroad car to St. Lawrence University, at Canton, New York. There she attended the dedication of the new science building Hepburn Hall, at one side of the entrance of which was a bas-relief of the English atomic scientist John Dalton, and on the other side, one of Marie Curie.

Her two days at Canton were busy. The first night a dinner was held in Canton's Hotel Harrington, to which a hundred of the town's leading citizens were invited. To make it easier for Marie, Owen Young, her escort, took her to a side table and then brought over one or two persons at a time to meet her.

One of these was a prominent North Country surgeon, Dr. Grant C. Madill, who had been using radium at Hepburn Hospital. The talk between him and Marie immediately became technical. The other guests watched with wonder when the doctor took a pencil from his pocket and started making diagrams or formulas on the back of his menu. Then they saw her hold up her hands, withered and scarred by radium burns, for him to examine. "A frail little lady, white-haired, but with eyes of youth," one of the guests described her later.

It was a tradition of St. Lawrence for the young students to serenade their sweethearts. The object of the youths' attention on this evening was Marie. They marched up Main Street six hundred strong, stopped in front of the home of the university president, Mr. Sykes, where she was staying, and then Marie heard their strong young voices singing out their college songs for her ears.

The next day, wearing her black silk robe with the velvet facings of her previous visit, she was given the honorary degree of doctor of science. Part of the dedication ceremonies for the new science building was the planting of a small spruce tree. President Sykes handed her a small iron shovel with which she was to turn up the first shovel of soil. A mischievous light came into her eyes. She shook her head at the president and grasped a full-sized workman's spade, which she used to cast a large shovelful of soil upon the roots of the tree as it was set in place. The spectators gasped with delight, while Owen Young, falling into the spirit of the occasion, seized the small shovel, ridiculously tiny for a man of his size, to heap on a few more tablespoons of dirt.

Two and a half years later she returned to Poland for the purpose of inaugurating the Radium Institute of Warsaw, for whose creation she was largely responsible. Warsaw had never seemed lovelier to her. One morning she slipped away and took a walk by herself along the Vistula River, bluish-green near at hand but made bluer far off by the reflection of the sky. In a letter to Ève about this stroll, she quoted an old Cracow song: *This Polish water has within itself such a charm that those who are taken by it will love it even until the grave.* She must have guessed at the time that this would be her last visit to her native land.

CHAPTER THIRTEEN

# End of the Journey

THAT SOMETHING AS POWERFUL AS radium could also be very dangerous was known from the first days of its discovery, but many facts about the hazards of radium rays were not found out until years later.

Between 1915 and 1930, before some of these damaging effects were known, radium water treatment was popular for arthritis and high blood pressure. This stopped abruptly when scientists learned that radium, taken internally, could prove a very lethal poison. Cases of malignant tumors and bone diseases were traced to those "radium waters."

Around 1924, the US Radium Corporation, in Orange, New Jersey, had employed forty-two women to paint watch dials with a luminous paint composed of radium and zinc sulfide. They had the habit of smoothing out the end of their brushes with their lips, thus taking into their system a very minute quantity of radium. One of these women died in 1925 of what was diagnosed as anemia. A doctor named Harrison S. Martland became suspicious and had all the women examined. He found that without exception they had radium poisoning. Over the years one after another died, the last in 1958—all victims of man's ignorance of the power of this strange substance that could both heal and destroy.

In the early days of radium's commercial development, there were many other radium workers and patients who suffered from the rays.

The situation is quite different today when the most painstaking precautions are taken. Accidents do happen occasionally. In 1951 a tiny platinum capsule of radium exploded at an electronics factory in the Midwest, and before it was discovered, the workmen had tracked the radium dust through the building. Immediately the building was decontaminated and held unfit for use until inspectors agreed there was no trace of radioactivity left.

Modern radiologists are appalled at the risks that the Curies took, particularly during their last months in the shed when they worked daily with highly radioactive materials. After their death, Claude Regaud, director of the Radium Institute Hospital, attributed the unusual fatigue felt by both Marie and Pierre at that time to internal lesions resulting from their exposure to radium and radium emanation.

Marie took further chances in the war years when she handled X-ray equipment every day. Nor was she any more sensible during the period of her directorship at the institute. Blood tests were required for all her helpers, though Marie wouldn't take them herself. When she was finally persuaded to do so, her blood content was found abnormal. This was not surprising after more than thirty years working with radium, and she wouldn't take it seriously.

She was full of plans for the future in that January of 1934. She had decided to have a villa built at Sceaux. The next fall she would move to a modern apartment. At Eastertime, when Bronya came from Poland to pay her a visit, she took her sister on a motor trip to the Mediterranean.

The trip proved too much for her. By the time they arrived, she had caught a bad chill, and Bronya had to put her to bed. It didn't turn out to be much of a vacation for either of them.

When she was a little better, she returned to Paris and her work. The chills and fever kept recurring, and she simply could not go to the institute every day. On days she wasn't strong enough to leave the apartment, she stayed home and worked on her book. It was a long one, and it was called *Radioactivity*.

In its pages she summed up in her clear, lucid style all that was now known on this topic. It was thirty-four years since she and Pierre had isolated radium salts and thirty-two years since Rutherford and Soddy had announced their theory of disintegration of radioactive substances. In this period tremendous strides had been made in physical science, and all of it stemmed from the discovery of elements that sent out rays that could not be altered by heat or cold or chemical treatment.

Many new radioactive elements had been discovered, and their relationship showed the pattern of a family tree with the two main branches of uranium and thorium, from which sprang a host of descendants.

As uranium disintegrated, it formed ionium, from which came radium; and as radium disintegrated, it formed the emanation gas radon, from which came radium A, B, C, and so on. Polonium too was a derivative of uranium, through the intermediary stage of radium. André Debierne's actinium was apparently another branch of the uranium family.

Protactinium was the immediate parent of actinium, and radioactinium was a descendant.

In the same way, mesothorium I was a derivative of thorium and was itself a source of radiothorium.

All of these new elements had now been analyzed, given an atomic weight, and assigned their place on the periodic table, and it was known how long it would take for each to disintegrate. Some of them, like uranium and thorium, were found to have a very long life, having survived for many years in the ores that contained them.

In comparison, ionium, with its half-life of 83,000 years, as well as radium, with its half-life of 1,620 years, were considered much younger indeed. At the other extreme were some of the gaseous derivatives of both families, with a half-life of only a few minutes or even a few seconds.

The composition of rays of radium and these other products had long since ceased to be a mystery. They were of three kinds, named by Ernest Rutherford, the New Zealander: *alpha*, *beta*, and *gamma* rays.

The alpha and beta rays could be deflected by a magnet, proving that they were not like light but were made up of tiny particles of matter. Without a magnet the three rays made one single beam. With the magnet, they were three different rays, the alpha ones slightly turned aside, the beta rays more markedly so, and the gamma rays unaffected by the magnet.

The gamma rays were similar in character to X-rays, and they were the ones used to treat cancer and penetrate thick metals.

The alpha rays were actually atoms of the element helium, shorn of their electrons, as Ernest Rutherford had discovered, and this was the first example of one element transforming itself into another.

He had tested these alpha particles by directing them against a thin sheet of gold leaf. Amazingly most of the alpha particles had breezed through the gold easily, though some had bounced back as if they had hit something solid.

Since gold is a heavy element, with its atoms packed closely together, this pointed to the conclusion that an atom was not as solid as many had thought, that there were empty spaces in it where most of the alpha particles passed without difficulty. The ones that bounced back had done so because they happened to hit the nucleus of the atom. The nucleus contained positively charged particles, given the name of protons.

# END OF THE JOURNEY

The beta rays were made up of infinitely smaller particles that turned out to be negatively charged electrons, first discovered by J.J. Thompson of the Cavendish Laboratories. Robert Andrews Millikan, of the University of Chicago, invented an electrical balance with which he was able to determine that the weight of an electron was some 1,850 times less than that of an atom of hydrogen, the lightest element known. It was Ernest Rutherford again who, from a study of alpha and beta rays, gave the world the first picture of the structure of the atom as a central nucleus with electrons revolving around it.

Other revelations had followed. Why was it that so many atoms had atomic weights in fractions—that of radium, for instance, being not quite 226—instead of in round numbers? The English scientist William Crookes back in 1886 expressed the belief that most atoms of the element calcium had an atomic weight of 40, but that there were a few with a weight of 39 or 41, and still fewer with a weight of 38 and 42. Frederick Soddy called such nonconforming atoms *isotopes*—and created another upheaval in the scientific world.

Ernest Orlando Lawrence, of Canton, South Dakota, then came out with the statement that even hydrogen had an isotope atom, heavier than the average hydrogen atom, and he named these heavy atoms *deuterium*. Harold Urey, a Columbia University scientist, proved this in 1931, finding that deuterium atoms occur in hydrogen, one part in 5,000, and for this he won the Nobel Prize in 1934. Later, great factories would be set up to isolate the heavy hydrogen atoms from ordinary atoms, and this substance, as heavy water, would be invaluable in atomic energy plants.

In 1932 James Chadwick, an English scientist, added another feature to the structure of the atom when he announced the existence of neutrons—particles that were neither negatively nor positively charged—that together with the positively charged protons made up the nucleus of many atoms. By this discovery he explained the different weights of isotopes: an ordinary hydrogen atom has one proton in its nucleus and one electron revolving in its outer space; deuterium, its isotope, has one proton and one neutron as well, in its nucleus.

Many other scientists from different parts of the world were making contributions to the new science, among them Niels Bohr, in Denmark, and Otto Hahn, who discovered the radio element mesothorium and, with Lise Meitner, radiothorium and protactinium in the Kaiser Wilhelm Institute of Berlin, and B.B. Boltwood, discoverer of ionium. Later, during the Second World War, Hahn and Lise Meitner, as exiles from Germany, would split the uranium atom, thus pointing the way to the production of atomic bombs. But this Marie would not live to see.

While all this exciting research work was being done, political events were leading the world slowly away from the conception of the Committee on Intellectual Cooperation for free interchange of information between scientists of all nations, which Marie believed in with all her heart.

In 1933 Adolf Hitler was made Chancellor of Germany. Albert Einstein was visiting professor of the California Institute of Technology at the time. He did not go back to Germany, and in this year of 1934, his property was confiscated by the Nazi government, and he was deprived of German citizenship on the grounds that he was a Jew. The policy that would cause Nazi Germany to lose its most distinguished scientists had already begun.

While Marie followed the researches of other scientists carefully, her own studies continued to be mainly with the properties of her pets, radium and polonium—improving old methods of analysis or seeing how these two could be useful in the discoveries of others. The English scientist A.S. Russell commented that after the isolation of radium metal in 1911, her attitude became protective rather than creative.

In 1924, for example, certain Russian scientists put up a claim that polonium decayed at different rates in different parts of the USSR. Indignantly, Marie flew to the defense of her element to prove it was impossible for such variations to take place. Her last researches were with S. Rosenblum, the young Jewish scientist in whose career she had interested Einstein, on alpha particles of varying initial velocity emitted by different products of the actinium disintegration series.

One chapter of Marie Curie's book on radioactivity was devoted to the work of two people who were very close to her, her daughter Irène and her son-in-law Frédéric, who at the time of their marriage had adopted "Joliot-Curie" as their last name. They had branched off from Marie's own research and the year before had created artificial isotopes that were themselves radioactive. The full implication of this discovery was not yet known, but Marie foresaw, as they did, that such radioactive isotopes, far less costly than radium, could be used to supplement and, in many cases, replace radium in the treatment of cancer. The following year they would receive the Nobel Prize for their achievement, a triumph that Marie Curie would not live to celebrate with them.

The Joliot-Curies often came to lunch on days when Marie was home. Sometimes they brought their little daughter, seven-year-old Hélène. Nothing pleased Marie more than the company of her grandchild.

She reported for her last day's work at the institute laboratory one day in May of 1934, but the fever returned, and she departed about the middle of the afternoon, with one final glance at the roses she had planted in the garden. She could not get out of bed after that and, with unusual docility, allowed Ève to call in doctors to examine her.

They could not be sure what was wrong, for she did not have the symptoms of any known disease. As she grew weaker, they recommended that she be sent to the sanatorium of Sancellemoz in the French Alps. Ève and a nurse took care of her in this difficult journey. It was thought that the pure mountain air would ease her condition. The new X-ray photographs there revealed that her lungs were not attacked, as the Paris doctors had suspected, but her fever remained consistently high, and the white corpuscles in her blood increased. The physicians at the sanatorium diagnosed her malady as pernicious anemia.

Irène and Frédéric came from Paris to join her on the 2nd of July. The next day, as she always did, she insisted on reading the thermometer that gave her temperature. She smiled with delight, for she noted that the fever had diminished. The mountain air had been good for her after all. She was going to get well. . . .

The others were not so hopeful. The lowering of a temperature is a common thing just before the end. They sat at her bedside for hours, holding her cold hands in theirs. In her delirium she stared fixedly at a cup of tea and asked whether it was made with radium or with mesothorium. The next day, July 4, 1934, was her last.

Only later did science give a verdict. The abnormal symptoms, the blood tests that gave results unlike any other known case of pernicious anemia, indicated the truth. Marie Curie had finally succumbed to radium.

She was buried by her family and close friends at Sceaux on July 6. On her grave her brother, Dr. Joseph Sklodovska, sprinkled soil that he had brought from Poland, combined with soil of France—symbols of her two countries. Her tomb was piled high with the flowers she loved so well, but there were no speeches. That was the way she would have preferred it.

MORE BOOKS FROM THE GOOD AND THE BEAUTIFUL LIBRARY

*Understood Betsy*
By Dorothy Canfield Fisher

*The Threatening Fog*
By Leon Ware

*Victory at Bear Cove—A Story of Alaska*
By Elsa Pedersen

*Treasure for Debby*
By Amy Wentworth Stone

GOODANDBEAUTIFUL.COM